Deepen Your Mind

Deepen Your Mind

大約十年前的某一天，我正徜徉在網際網路的世界裡，忽然一個名叫 "TinEye" 的影像搜尋引擎網站映入我的眼簾。我滿懷憧憬地在那個網站中上傳了一幅圖片，它很快搜索並傳回了許多這幅圖片在網際網路中不同 URL 上的結果。我接著嘗試上傳了另一幅圖片，一會兒它又傳回了許多近似這幅圖片的結果，很顯然，結果中的很多圖片是在同一幅影像上修改的。面對如此準確和令人驚豔的結果，我不禁腦洞大開、浮想聯翩，構思著一個個可以運用該技術實現的奇思妙想。猛然間，我覺得心中產生了一股強大的力量 -- 我要弄清楚它背後的技術原理。

為了徹底弄清楚這種別圖像搜尋引擎的技術原理，我反覆尋找和閱讀當時網際網路上甚為缺乏的相關資料，但收效甚微。直到後來，我遇到了一個叫作 LIRE 的開放原始碼專案，它讓我初步了解影像搜尋引擎的技術原理。但是在實際應用中，LIRE 的效果並不是太好。為了解決這個問題，我又找到「深度學習」這個強有力的幫手。在探索原理的過程中，我發現幾乎找不到一本介紹影像搜尋引擎基本原理和實現的中文書，這也成了本書誕生的緣由。

以內容為基礎的影像檢索技術自 20 世紀 90 年代提出以來，獲得了迅速的發展。研究人員提出了不同的理論和方法，其中具有代表性的是 SIFT、詞袋模型、向量量化、倒排索引、局部敏感雜湊、旋積神經網路，等等。與此同時，產業界也推出了許多實用的影像搜尋引擎，例如 TinEye、Google 影像搜索、百度影像搜索和以淘寶為代表的垂直領域影像搜尋引擎。但是到目前為止，此項技術還遠未完全成熟，還有許多問題需要解決，改進和加強的空間還很大。搜索的結果和使用者的期望還有一些距離，存在一定的影像語義鴻溝。這也是從事這項技術研究與開發的人員不斷進步的源動力。

希望本書的出版能夠在某種程度上緩解影像搜尋引擎資料稀少的現狀，並能夠吸引和幫助更多的技術人員關注並研究影像檢索技術。

明恒毅

得益於以內容為基礎的影像檢索技術的發展，近十年來網際網路業界湧現出一些以 TinEye 影像搜索、淘寶影像搜索為代表的通用和垂直領域影像搜尋引擎。這些影像搜尋引擎改變了以往單一的關鍵字檢索方式，相當大地滿足了人們日益多樣的影像檢索需求。作者在研究影像搜尋引擎的過程中發現，目前尚無一本系統論述影像搜尋引擎原理與實現的中文書籍，因此產生了撰寫本書的想法。

本書內容共分為 5 章。

- 第 1 章由文字搜尋引擎的原理講起，逐步抽象出搜尋引擎的一般結構，帶領讀者由文字搜索過渡到影像搜索。

- 第 2 ～ 3 章分別按照傳統人工設計和深度學習兩種方式，對影像特徵分析的相關理論和方法說明。

- 第 4 章詳述了影像特徵索引和檢索的相關理論和方法。

上述每一章都在說明相關理論和方法的同時，使用以 Java 語言為基礎的實現程式和詳實的程式註釋對理論和方法進行複述。力求讓讀者不但能夠了解深奧的理論知識，而且能將理論轉為實際可執行的程式。

第 5 章會帶領讀者從零開始逐步建置一個以深度學習為基礎的 Web 影像搜尋引擎，讓讀者能夠更透徹地了解影像檢索的理論，並具有獨立實現一個線上影像搜尋引擎的能力。

影像搜尋引擎技術涵蓋知識面廣，目前尚在不斷發展中，由於作者水準所限，書中難免存在錯誤和不足之處，歡迎各位讀者批評指正。回饋意見和建議可以透過加入本書 QQ 群（743328332）進行溝通交流，或致信電子郵件 imgsearch@126.com，我將不勝感激。

在這裡，我要感謝人民郵電出版社編輯張爽的邀請，透過寫作此書，我感受到技術寫作的不易與樂趣，也獲得一次難得的提升能力的機會。還要感謝父母妻兒對我的支援和了解，以及生活上的照顧，正是有了他們的支援，才能讓我能夠心無旁騖、安心寫作。

Contents 目錄

01 從文字搜索到影像搜索

02 傳統影像特徵分析

03 深度學習影像特徵分析

04 影像特徵索引與檢索

05 建置一個以深度學習為基礎的 Web 影像搜尋引擎

從文字搜索
到影像搜索

1.1 文字搜尋引擎的發展[1]

1990 年，加拿大麥吉爾大學的 Alan Emtage 等學生開發了一個名叫 Archie 的系統。該系統透過定期搜集分析散落在各個 FTP 伺服器上的檔案名稱清單，並將之索引，以供使用者進行檔案查詢。雖然該系統誕生在 WWW 的出現之前，索引的內容也不是現代搜尋引擎索引的網頁資訊，但它採用了與現代搜尋引擎相同的技術原理，因此被公認為現代搜尋引擎的鼻祖。

1991 年，明尼蘇達大學的學生 Mark McCahill 設計了一種用戶端 / 伺服器協定 Gopher，用於在網際網路上傳輸、分享文件。之後產生了 Veronica、Jughead 等類似 Archie，但執行於 Gopher 協定之上的搜索工具。

---

1 Michael Busby. Learn Google: Wordware Publishing, Inc., 2003

同一時期，英國電腦科學家 Tim.Berners.Lee 提出了將超文字和 Internet 相結合的想法，並將之稱為 WWW（World Wide Web）。隨後，他創造了第一個 WWW 的網頁，以及瀏覽器和伺服器。1991 年，他將該專案公之於眾。自此，WWW 成為了 Internet 的主流，全球進入豐富多彩的 WWW 時代。搜尋引擎也逐步從 FTP、Gopher 過渡到 WWW，並進一步演進。

1993 年，麻省理工學院的學生 Matthew Gray 開發了第一個 WWWspider 程式 WWW Wanderer，它可以沿著網頁間的超連結關係一個一個存取。起初，WWW Wanderer 只是用來統計網際網路上的伺服器數量，後來加入了捕捉 URL 的功能。雖然它功能比較簡單，但它為後來搜尋引擎的發展提供了寶貴的思維參考。這一構思激勵了許多研究開發者在此基礎上進行進一步改進和擴充，並將 spider 程式抓取的資訊用於索引建置。我們今天在開發一個網站或做搜尋引擎最佳化時所用到的 robot.txt 檔案，正是告訴 spider 程式可以爬取網站的哪些部分，不可以爬取哪些部分的一份協定。同年，英國 Nexor 公司的 Martin Koster 開發了 Aliweb。它採用使用者主動傳輸網路頁簡介資訊，而非程式抓取的方式建立連結索引。是否使用 robot、spider 擷取資訊也形成了搜尋引擎發展過程中的兩大分支，前者發展為今天真正意義上的搜尋引擎，後者發展為曾經風靡一時，能夠提供分類目錄瀏覽和查詢的入口網站。

1994 年可以說是搜尋引擎發展史上里程碑的一年。華盛頓大學的學生 Brain Pinkerton 開發了第一個能夠提供全文檢索的搜尋引擎 WebCrawler。而在此之前，搜尋引擎只能夠提供 URL 或人工摘要的檢索。自此，全文檢索技術成為搜尋引擎的標準配備。這一年，史丹佛大學的楊致遠和 David Filo 建立了大家熟知的 Yahoo，使資訊搜索的概念深入人心，但其索引資料都是人工輸入的，雖能提供搜索服務，但並不能稱之為真正的搜尋引擎；卡內基美隆大學的 Michael Maldin 推出了 Lycos，它提供了搜索結果的相關性排序和網頁自動摘要，以及字首比對和字元近似，是搜尋引擎的又一歷史性進步；搜尋引擎公司 Infoseek 成立，在其隨後的發展中，它第一次允許站長傳輸網路址給搜尋引擎，並將「千人成本」（Cost Per Thousand Impressions，CPM）廣告模式引入搜尋引擎。

1995 年，一種全新類型的搜尋引擎 -- 元搜尋引擎誕生了，它是由華盛頓大學的學生 Eric Selburg 和 Oren Etizioni 開發的 MetaCrawler。元搜尋引擎採用將使用者的查詢請求分發給多個預設的獨立搜尋引擎的方式，並統一傳回查詢結果。但是由於各獨立搜尋引擎搜索結果的評分機制並不相同，常常傳回一些不相干的結果，精準性並不如獨立搜尋引擎好，因此元搜尋引擎始終沒有發展起來。

同一年，DEC 公司開發了第一個支援自然語言搜尋及布林運算式（如 AND、OR、NOT 等）進階搜索功能的 AltaVista。它還提供了新聞群組搜索、圖片搜索等具有劃時代意義的功能。

1998 年，史丹佛大學的學生 Larry Page 和 Sergey Brin 創立了 Google（Google）-- 一個日後影響世界的搜尋引擎。Google 採用了 PageRank（網頁排名）的演算法，根據網頁間的超連結關係來計算網頁的重要性。該演算法相當大地加強了搜索結果的相關性，使其後來居上，幾乎壟斷了全球搜尋引擎市場。

1.2 文字搜尋引擎的結構與實現

目前，以文字資訊為基礎的搜尋引擎雖然還有一定的提升空間，但其工作原理已經相對穩定，基本結構也已趨於成熟。文字搜尋引擎基本可以分為抓取部分、前置處理部分、索引部分、搜索部分以及使用者介面，如圖 1-1 所示。

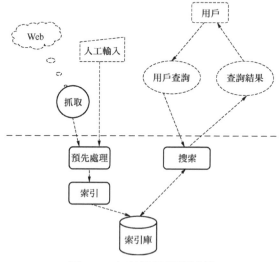

圖 1-1 文字搜尋引擎結構

由於抓取部分不是本書所討論的內容，故不做詳細介紹。下面主要介紹文字資料前置處理、索引及搜索。

1.2.1 文字前置處理

網頁自動尋檢程式（Spider）抓取的資料在進行某種程度的前置處理之後才能用於索引的建立。文字資料前置處理主要是為了分析詞語而進行的文字分析，而文字分析又可分為分詞、語言處理等過程。

■1 分詞

文字分詞過程通常分為三步驟：第一步，將文字分為一個個單獨單字；第二步，去除標點符號；第三步，去除停止詞（Stop words）。停止詞是語言中最普通的一些單字，它們的使用頻率很高，但又沒有特殊意義，一般情況下不會作為搜索關鍵字。為了減小索引的大小，一般將這種單字直接去除。為方便讀者了解，下面舉例說明，如圖 1-2 所示。

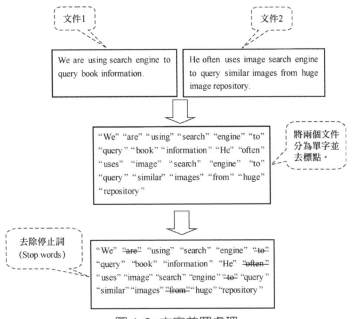

圖 1-2　文字前置處理

■2 語言處理

語言處理主要對分詞產生的詞元進行對應語言的處理。以英文為例：首先將詞元變為小寫，然後對單字進行縮減。縮減過程主要有兩種，一種被稱為詞幹分析（Stemming），另一種被稱為詞形還原（Lemmatization）。詞幹分析是取出詞的詞幹或詞根，詞形還原是把某種語言的詞彙還原為一般形式。兩者依次進行相關語言處理，例如將

books 縮減為 book（去除複數形式），將 tional 縮減為 tion（去除形容詞尾字母）。詞幹分析採用某種固定的演算法進行縮減。詞形還原通常使用字典的方式進行縮減，縮減時直接查詢字典，例如將 reading 縮減為 read（字典中存在 reading 到 read 的對應關係）。詞幹分析和詞形還原有時會有交集，同一個詞，使用兩種方式都會獲得同樣的縮減。接上面的舉例，繼續說明，如圖 1-3 所示。

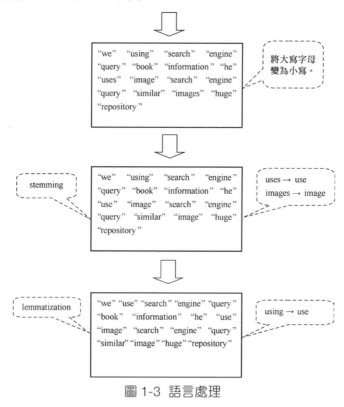

圖 1-3 語言處理

1.2.2 建立索引

經過文字分析後，獲得的結果稱為詞（Term），我們利用它建立索引。首先使用獲得的詞建立一個字典，然後對字典按字母順序進行排序，最後合併相同的詞，形成文件倒排表（Posting List），實際過程如下。

① 使用詞產生字典，如表 1-1 所示

表 1-1 使用詞產生字典

詞	文件 ID	詞	文件 ID
we	1	image	2
use	1	search	2
search	1	engine	2
engine	1	query	2
query	1	similar	2
book	1	image	2
information	1	huge	2
he	2	repository	2
use	2		

② 對字典按字母順序排序，如表 1-2 所示

表 1-2 對字典按字母順序排序

詞	文件 ID	詞	文件 ID
book	1	query	2
engine	1	repository	2
engine	2	search	1
he	2	search	2
huge	2	similar	2
image	2	use	1
image	2	use	2
information	1	we	1
query	1		

3 合併相同的詞，形成文件倒排鏈結串列

在文件倒排表中，有幾個概念需要解釋一下。文件頻率（Document Frequency）表示共有多少個文件包含這個詞。詞頻率（Term Frequency），表示這個文件中包含此詞的個數。在圖 1-4 中，左邊是按字母順序排序的字典合併相同詞，並統計出該詞在文件中出現次數的結果。中間和右邊是文件 1 和文件 2 中包含某個詞的次數 -- 詞頻率。它們之間是用鏈結串列的形式串起來的，又因為是根據詞的值來尋找相關文件的，而非在文件中尋找相關的值，和正常順序是相反的，故稱其為文件倒排鏈結串列或倒排索引。

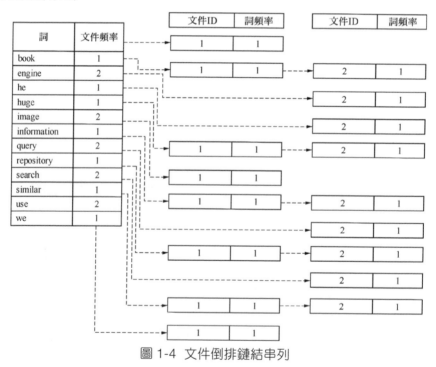

圖 1-4 文件倒排鏈結串列

至此，索引已經建置好了。根據以上的文件倒排鏈結串列，我們就能使用關鍵字來查到對應的文件了。

1.2.3 對索引進行搜索

上面我們已經可以尋找到包含關鍵字的相關文件了，但它還不能滿足實際搜索的要求。如果結果只有幾個，當然沒有問題，全部顯示就是了。但在實際應用中，搜尋引擎需要傳回幾十萬，甚至百萬、千萬級的結果。我們怎樣才能將最相關的文件顯示在最前面呢？這也是下面需要探討的問題。

▉ 使用者輸入查詢敘述

目前，搜尋引擎均提供自然語言搜尋以及布林運算式進階搜索，所以查詢敘述也是遵循一定的語法結構。例如我們可以輸入查詢敘述 "search AND using NOT image"，它搜索包含 search 和 using 但不包含 image 的文件。

▉ 對查詢敘述進行詞法分析、語法分析、語言處理

詞法分析用來分析查詢詞以及布林關鍵字，上面的查詢敘述分析出的查詢詞為 search using image，布林關鍵字是 AND 和 NOT。語法分析會將詞法分析的分析結果產生一棵語法樹。上例形成的語法樹如圖 1-5 所示。

語言處理與建立索引時的語言處理過程幾乎相同。如圖 1-6 所示，上例中的 using 將轉為 use。

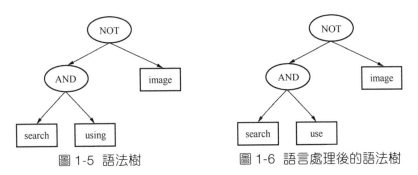

圖 1-5 語法樹　　　　　　圖 1-6 語言處理後的語法樹

❸ 搜索索引，傳回符合上述語法樹的結果

首先，在反向索引中分別找出包含 search、use 和 image 的文件鏈結串列。然後，將包含 search 和 use 的文件鏈結串列合併，獲得既包含 search，又包含 use 的文件鏈結串列。接著，在上一步驟的結果中去除包含 image 的文件鏈結串列，最後的文件鏈結串列就是符合上述語法樹的結果。

❹ 對結果進行相關性排序

雖然在上一步驟中我們獲得了想要尋找的文件，但這些文件並未按照與查詢敘述的相關性進行排序，並不是我們最後想要的結果。那怎樣才能將尋找結果按相關性進行排序呢？首先，把查詢敘述也視為一個由許多片語組成的短小文件，那麼查詢敘述與對應文件的相關性問題就轉變成了文件之間的相關性問題。毫無疑問，文件主題近似程度高的，其相關性必然強；文件主題近似程度低的，其相關性必然弱。那進一步思考一下，什麼又是決定文件間相關性的主要因素？想必讀者都讀過或寫過論文，是否留意到每篇論文都有「關鍵字」這一項？「關鍵字」是能夠反映論文主題的詞或片語。也就是説，論文中每個詞對論文主題思維的表達程度是不相同的。換個説法，文件中的每個詞對其主題思維表達的加權是不同的，正是這些不同加權的詞組成了文件的主題。

有兩個主要因素會影響一個詞在文件中的重要性。一是詞頻率（Term Frequency，tf），表示一個詞在此文件中出現的次數，它的值越大，説明這個詞越重要。二是文件頻率（Document Frequency，df），表示多少文件中包含這個詞，它的值越大，説明這個詞越不重要。一個詞的加權可以使用式（1-1）進行計算：

$$W_{t,d} = tf_{t,d} \times \log\left(\frac{n}{df_t}\right) \tag{1-1}$$

在上述公式中，$W_{t,d}$ 表示詞 t 在文件 d 中的加權，$tf_{t,d}$ 表示詞 t 在文件 d 中出現的頻率，n 表示文件的總數，df_t 表示包含詞 t 的文件數量。

怎樣才能度量文件的相似度呢？第一步，把文件中每個詞的加權組成一個向量，$DocumentVector=\{weight1, weight2, \cdots, weightN\}$。把查詢敘述也看作一個簡單的文件，將其中的每個詞的加權也組成一個向量，$QueryVector=\{weight1, weight2, \cdots, weightN\}$。第二步，將所有查詢出的文件向量和查詢敘述向量取聯集，用聯集元素的個數 N 統一各向量長度，如果一個文件中不包含某個詞，那麼該詞的加權為 0。第三步，把所有統一後的向量放到一個 N 維空間中，每個詞是一維，如圖 1-7 所示。

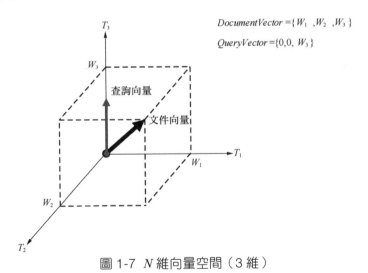

圖 1-7 N 維向量空間（3 維）

如圖 1-8 所示，檔案向量間存在一定的夾角。我們可以透過計算夾角餘弦值的方法來表示它們之間的相似程度。因為夾角越小，餘弦值越大，也就是說文件向量夾角的餘弦值越接近，文件也越相近。

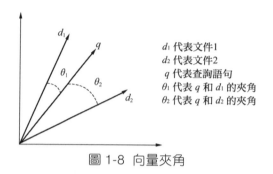

圖 1-8 向量夾角

相關性的計算公式如下：

$$Similarity(q, d) = \cos(q, d) = \frac{\vec{V_q}\vec{V_d}}{|\vec{V_q}||\vec{V_d}|} = \frac{\sum_{i=1}^{n} w_{i,q}w_{i,d}}{\sqrt{\sum_{i=1}^{n} w_{i,q}^{2}} \sqrt{\sum_{i=1}^{n} w_{i,d}^{2}}} \qquad (1\text{-}2)$$

整理一下上述的索引、搜索過程，如圖 1-9 所示。

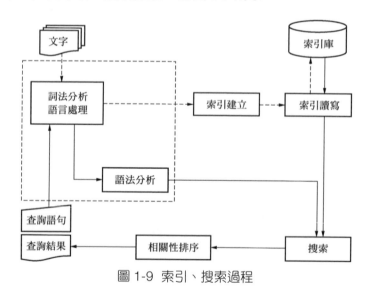

圖 1-9 索引、搜索過程

詞法分析和語言處理過程將一系列文字轉化為許多個詞，然後索引建立過程將這些詞產生詞典和倒排索引，索引寫程式將其寫入索引函數庫。當使用者輸入查詢敘述進行搜索時，首先進行詞法分析和語言處理，將查詢敘述分解成一系列詞，而後將其輸入語法分析過程產生查詢語法樹。索引讀程式將反向索引表由索引函數庫讀取記憶體，搜索過程在反向索引表中尋找與查詢語法樹中每個詞一致的文件鏈結串列，並進行對應的布林運算，獲得結果文件集。將結果文件集與查詢敘述的相關性進行排序，並將產生結果傳回給使用者。

1.3 搜尋引擎的一般結構

在學習了文字搜尋引擎之後，我們是否可以從文字搜尋引擎抽象出搜尋引擎的一般結構呢？根據一般的抽象方法，我們可以把事物非關鍵性的特徵剝離出來，而只保留其最為本質的特徵。對於現有技術條件下的搜尋引擎，必須事先產生索引函數庫，再在其上進行搜索查詢。如圖 1-10 所示，首先需要對輸入資料進行一定的前置處理，以使我們可以進一步分析。接下來，把文字搜尋引擎的詞法、語法分析等語言處理階段抽象為對輸入資料的特徵分析，一個個分析出來的詞就是組成一個文件特徵向量的基本元素，反向索引函數庫就是特徵和文件對應關係的集合。對於查詢資料，我們也要取出其特徵，然後計算它的特徵向量與索引函數庫中所有特徵向量的相似度，最後傳回規定數量的相似結果。

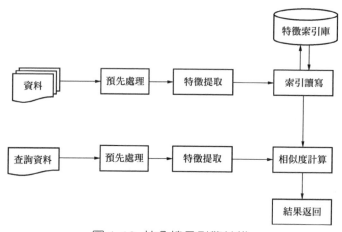

圖 1-10 抽象搜尋引擎結構

1.4 從文字到影像

隨著網際網路的發展和網路頻寬的改善，WWW 上的資訊逐漸由純文字過渡到文字和影像相結合，甚至有些網站（例如 Flickr 和 Pinterest）發佈的資訊幾乎都是影像。尋找文字資訊是傳統搜尋引擎的強項，但對於影像資訊，很多搜尋引擎無能為力。面對使用者強烈的需求，很多網際網路公司開始在自己的搜尋引擎中增加影像搜索的選項。

下面來觀察一個網頁的原始檔案：

```html
<html>
<head>
    <title>貓</title>
</head>
<body>
    <img src="/images/animal/cat.jpg" alt=" 可愛的小貓 " />
    <div class="introduction" label-module="t1">貓身體靈活，樣子招人喜愛。
```

```
</div>
</body>
</html>
```

使用者一看就知道這是一個介紹貓的網頁。html 檔案的 title 是貓，該檔案中還有一個小貓的圖片，其路徑是 /images/animal/cat.jpg，並且有一個 Alt 標籤說明了圖片的內容，圖片下面還有一段貓的簡介。我們是否可以利用這些內容來索引和搜索影像呢？答案是一定的。

最初，Altavista、Lycos 等搜尋引擎正是利用影像的檔案名稱、路徑名稱，影像周圍的文字以及 Alt 標籤中的註釋索引和搜索相關影像的。從本質上來說，這樣的影像搜尋引擎其實還是以文字搜尋引擎為基礎的。有時影像週邊的這些文字資訊和影像並沒有關係，會造成搜索出來的部分影像結果和查詢關鍵字並不一致。為了避免這種缺陷，有些搜尋引擎採用人工的方式對影像進行標記索引。例如美國中北部教育技術聯盟開發的 Amazing Picture Machine，它事先由專人從事影像資訊的搜集、整理和標記，雖然人工標記確保了搜尋引擎的查準率，但是它限制了影像索引的規模，不可能有很好的查全率。

有時，影像的內容是很難用幾個關鍵字就能完整描述出來的。在某種情況下，無論是利用影像網頁相關文字資訊，還是人工標記文字說明，都很難做到較高的搜索準確度。1992 年，T. Kato 提出了以內容為基礎的影像檢索（CBIR）概念，它使用影像的顏色、形狀等資訊作為特徵建置索引用於影像檢索，也即我們通常所說的「以圖查圖」。以這一概念，IBM 開發了第一個商用為基礎的 CBIR 系統 QBIC（Query By Image Content），使用者只需輸入一幅草圖或影像便可以搜索出相似的影像。同一時期，很多公司也將這一技術引用搜尋引擎。哥倫比亞大學開發的 WebSEEK 系統，不僅提供了以關鍵字為基礎的影像搜索和按照影像類別目的主題瀏覽，還可以利用影像的顏色資訊進行以內容為基礎的影像搜索。Yahoo 的 ImageSurfer 也提供了使使用案例圖的顏色、形狀、紋理特徵以及它們的組合來進行以內容為基礎的影像搜索功能。隨著視覺技術的進步和發展，越來越多的搜尋引擎採用這一方式來進行影像搜索，並在此基礎上不斷演進。

曾經使用「以圖搜圖」方式進行過影像搜索的讀者可能都會有這樣的印象，這種影像搜索傳回結果的準確度常常不是太令人滿意。為此，很多視覺研究人員、影像技術開發者不斷提出新的影像特徵表示演算法。雖然準確率在一點點加強，但是並未根本性地解決這一問題。這究竟是什麼原因呢？原因就在於無論是影像的顏色、紋理、形狀這些全域資訊，還是後來的 SIFT 等局部影像資訊都是人為設計的強制寫入，還不能完整地表達人類對整幅影像內容的了解。那影像搜索的準確率還能加強嗎？隨著人工智慧，特別是神經網路理論和技術的發展，人們逐步找到了解決方案。

神經網路演算法起源於 1943 年的 MCP 類神經元模型，經過諸多科學家的努力，歷經跌宕起伏的發展，它逐步解決了發展中的問題，進入了新的快速發展階段。2006 年，Hinton 提出了訓練深層神經網路的新想法，也就是現在所說的深度學習。2012 年，Hinton 和他的學生 Alex 等人參加 ImageNet 影像識別比賽，利用深度學習理論建置的旋積神經網路（CNN）AlexNet 以 84.7% 的正確率一舉奪冠，並以相當大的優勢擊敗了使用人工設計特徵演算法獲得亞軍的選手。自此，在影像特徵分析方法上，深度學習的方法超過了許多傳統方法。很多影像搜尋引擎也引用了深度學習演算法，相當大地加強了影像搜索的準確率。

1.5 現有影像搜尋引擎介紹

目前很多網際網路公司都推出了影像搜尋引擎,並不斷改進,使其獲得
了日新月異的發展。下面選取幾個有代表性的影像搜尋引擎進行簡介。

1.5.1 Google 影像搜尋引擎

但凡提到搜尋引擎,就不可能繞過 Google。2001 年 7 月,Google 第
一次推出了影像搜索服務。最初,它只支援以文字資訊為基礎的影像
搜索,使用前面介紹的圖片名稱、路徑、Alt 標籤、圖片周圍的說明文
字等進行索引。2011 年 6 月,Google 在其影像搜索的首頁中加入了以
內容為基礎的影像搜索功能。隨著相關領域理論和技術的快速發展,
Google 不斷改進,搜索結果的準確度也不斷獲得提升。

使用者介面如圖 1-11 所示(為方便讀者了解,做了一定處理,下同),
搜索框中可以輸入所要查詢影像的相關詞彙,後面灰色的照相機按鈕用
於上傳影像或輸入圖片網址來查詢與之相似的許多結果。

圖 1-11 Google 影像搜尋引擎

圖 1-12 展示了使用 Google 影像搜尋引擎上傳一個香蕉圖片時傳回的結果。在傳回的頁面中，搜索框中顯示了我們剛剛上傳的圖片，並進行了影像到文字的解析，解析為 banana family。下面依次為對 banana family 的文字搜索結果和與上傳圖片外觀類似影像的展示。很顯然，Google 在影像搜索過程中，無論是以文字為基礎的，還是以內容為基礎的，都大量使用了人工智慧技術。

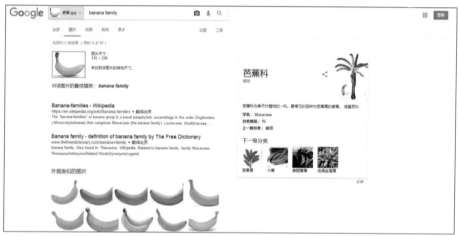

圖 1-12 Google 影像搜尋引擎查詢結果

1.5.2 百度影像搜尋引擎

百度在影像搜索領域和 Google 基本上走的是相同的路線。如圖 1-13 所示，百度具有與 Google 相同的搜索框和相機按鈕，不過百度將之稱為「百度識圖」。這次我們依然上傳那幅香蕉圖片檢視一下搜索結果。如圖 1-14 所示，百度識圖的傳回結果基本和 Google 類似，對圖片內容進行了解析，並且傳回與圖片內容相關的網頁資訊和相似圖片。百度的搜索結果反映了其在人工智慧領域取得的豐碩成果已可以和 Google 相匹敵。

圖 1-13 百度影像搜尋引擎

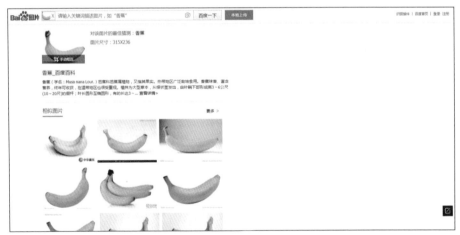

圖 1-14 百度影像搜尋引擎查詢結果

1.5.3 TinEye 影像搜尋引擎

TinEye 是一個純粹的反向影像搜尋引擎，可以説是利用已有影像搜索相似影像的鼻祖。TinEye 由加拿大 Idée 公司開發，於 2008 年 5 月正式上線，如圖 1-15 所示。它可以根據使用者填寫的影像 URL 或上傳影像進行搜索。TinEye 只專注於影像反向搜索,而不使用和影像相關的文字資訊。下面來看一下它的效果,見圖 1-16,可以看到它的傳回結果相較 Google 和百度少了很多,這也顯示 TinEye 在這一輪的影像人工智慧競賽中明顯落後了。

圖 1-15 TinEye 影像搜尋引擎

圖 1-16 TinEye 影像搜尋引擎查詢結果

1.5.4 淘寶影像搜尋引擎

淘寶的影像搜尋引擎同前述介紹的 3 個通用影像搜尋引擎並不十分相同。如圖 1-17 所示，它是一個垂直領域的影像搜尋引擎，索引了淘寶極大的商品函數庫資訊。它提供給使用者相似商品的查詢服務，以及依照圖片尋找商品的功能。我們依舊上傳那個香蕉圖片，獲得如圖 1-18 所示的結果。在搜索結果中，我們看到系統已經把它歸入「其他」類別目，顯然系統設計了一個粗略的分類器。頁面中「外觀相似寶貝」部分展示了淘寶商品函數庫中的外觀類似商品，不僅有實物香蕉，還有道具香蕉。

圖 1-17 淘寶影像搜尋引擎

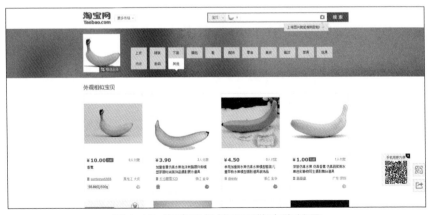

圖 1-18 淘寶影像搜尋引擎查詢結果

1.6 本章小結

本章沿著歷史的脈絡，由現代搜尋引擎的鼻祖 Archie 到如今壟斷全球市場的 Google，回顧了文字搜尋引擎的發展史，緊接著從原理上分文字前置處理、索引、搜索這 3 部分介紹了文字搜尋引擎的結構與實現。由此抽象出搜尋引擎的一般結構，並進一步將搜尋引擎從文字推導到影像，整理了它們的發展軌跡。最後以 Google、百度、TinEye、淘寶影像搜尋引擎為例，分析了它們各自的對話模式和技術特點。

傳統影像
特徵分析

2.1 人類怎樣取得和了解一幅影像

當可見光透過晶狀體射入人的眼光時，會在視網膜上成像。而視網膜上的不同感光細胞受到光源刺激後，會把所受刺激轉變成生物電訊號經由視神經傳入大腦，最後在大腦中形成一幅影像。

視網膜上存在視錐細胞和視桿細胞兩種感光細胞。視錐細胞光感感度低，只有當光源達到一定強度時，才能使其產生反應，但它具有分辨顏色的能力。經人類的實驗證實，視錐細胞中存在分別對峰值波長為 420nm、530nm、560nm（幾乎相當於藍、綠、紅三原色的波長）的可見光敏感的 3 大類細胞。視桿細胞不能感知顏色，卻對光源敏感度高，可以在光線很暗的情況下產生反應。我們常有夜晚能夠看到物體，卻不能分辨實際顏色的經歷，就是因為這個原因。

然而人類是怎樣對看到的影像進行了解的呢？根據成長經歷，我們常常遵循由外而內、由表及裡的方法觀察一個事物。我們首先獲知事物的顏色、大小、形狀、質地等外部特徵。一個小孩子都能分辨一個水果是蘋果還是香蕉，這是因為他親眼看到了這種水果，大人們又告訴了他蘋果是什麼顏色、何種形狀、有多大尺寸，而香蕉是什麼顏色、何種形狀、大小又是怎樣。這樣就使他在大腦中產生了蘋果、香蕉和各自顏色、形狀、大小的明確對應關係，如圖 2-1 所示。日後，他一看到這種顏色、形狀、大小的物體，他就能立即説出它是蘋果還是香蕉。

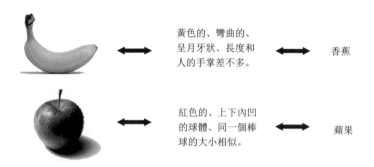

圖 2-1 影像與特徵以及名稱的對應

2.2 電腦怎樣取得和表示一幅影像

我們已經知道了人類如何獲知和了解一幅影像。那有讀者可能會問電腦取得和表示影像又是怎樣一個過程呢？和我們人類又有什麼區別呢？下面來詳細介紹。

電腦透過影像感測器取得物體的原始影像，它就好似我們人類的眼睛。根據組成元件的不同，影像感測器又分為金屬氧化物半導體元件（CMOS）和電荷耦合元件（CCD）兩種類型。雖然它們的組成元件不同，但所取得的原始模擬影像最後都會經過取樣、量化、數位影像處理等環節，最後輸出一幅數位化的影像。

2.2.1 取樣

影像在空間上的離散化，我們稱之為取樣。也就是在水平和垂直方向上，按照一定的空間間隔選取影像上的某些微小區域，這些區域稱為像素，其過程如圖 2-2 所示。

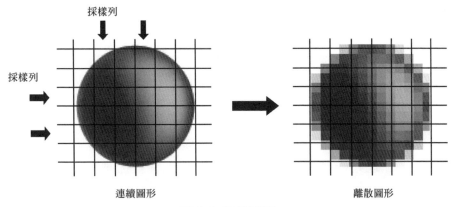

圖 2-2 取樣過程

「水平方向上的像素數 × 垂直方向上的像素數」，這種表示法稱為影像的解析度。例如一個影像的解析度是 800 像素 ×600 像素，那麼該影像每行有 800 個像素，每列有 600 個像素，總共由 800×600=480000 個像素組成。取樣的像素越密集，那麼它的解析度也就越高，影像也就越清晰。

2.2.2 量化

模擬影像在取樣之後，在空間上已經離散化為一系列像素，然而這時像素的灰階值還是由無窮個值組成的連續變化量。如圖 2-3 所示，我們把將這些像素各色彩分量的灰階值離散化的過程叫作量化。

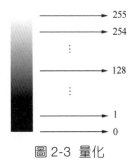

圖 2-3 量化

量化的過程會把這些連續的值轉化為一定數量的灰階值，這些灰階值的數量就決定了影像色彩或光線的豐富程度。例如設定影像每個色彩分量灰階值的數量為 256（2^8），也就把每個像素劃分為 0 ～ 255 個灰階，即量化為 8bit 色彩。

2.2.3 數位影像的儲存

模擬影像經過取樣、量化之後輸出為數位影像。該數位影像最後需要以一定的格式儲存在硬碟、隨身碟、SD 卡等存放裝置上。

數位影像按照組成方式的不同分為點陣圖和向量圖。點陣圖就是上述由像素組成的影像，每個像素包含了顏色、灰階、明暗、比較度等資訊。由於點陣圖是按照一定的解析度和量化位數取樣量化而成的，因此點陣圖一旦產生，像素的資訊就固定下來。當我們放大影像時，由於像素數量無法改變，放大倍數越高，影像就越不清晰。為了克服點陣圖這一缺點，一種新的影像格式 -- 向量圖應運而生。向量圖以數學向量代替位，儲存影像資訊，由按照數學公式計算產生的點、線組成。這一特性使圖形的縮放、移動、旋轉變得更為簡單，只需改變公式中的向量變數即可。點陣圖可以表現豐富的色彩，再現真實世界，但會產生縮放失真的現象。向量圖能夠無限縮放而不失真，但難以表現色彩豐富的逼真影像效果。向量圖主要由設計軟體產生，如 CorelDraw、FreeHand 和 AutoCAD。人們日常拍攝的數位照片，掃描的影像均是點陣圖的一種存在形式。本書所討論的影像均是點陣圖。

2.2.4 常用的點陣圖格式

常用的點陣圖格式有 BMP、JPEG、GIF 和 PNG 等。

BMP 全稱 Bitmap，它是 Windows 作業系統標準的影像檔格式。BMP 檔案格式在 Windows 3.0 以前與顯示裝置相關，稱為裝置相關點陣圖（DDB）；在 Windows 3.0 以後與顯示裝置無關，稱為裝置無關點陣圖（DIB）。它採用位對映儲存，未對圖像資料進行壓縮，故該格式的圖片會佔用更大的儲存空間。

JPEG 是聯合影像專家團隊的縮寫，它是第一個國際影像壓縮標準。為了解決圖片體積過大的問題，JPEG 採用失真壓縮方式去除了容錯的圖像資料，但由於壓縮演算法設計精妙，圖片依然保持了很高的品質，它還可以靈活地讓使用者在影像品質和體積之間做取捨。

GIF（Graphics Interchange Format）是 CompuServe 公司在 20 世紀 80 年代提出的檔案格式，全稱為影像交換格式。GIF 檔案格式採用 LZW 無失真壓縮演算法進一步縮小了體積，但最多只支援 256 種色彩。GIF 檔案格式不僅支援靜態圖片，還支援動畫格式。它有兩個版本，分別是 1987 年制定的 GIF87a 和 1989 年制定的 GIF89a，GIF89a 擴充了 GIF87a 的功能，提供了對透明色和多頁框動畫的支援。GIF 是一種需授權使用的專利格式，商業使用需要支付專利費，為此人們開發了另一種影像格式 -- 便攜網路圖形（Portable Network Graphics，PNG）。

PNG 格式採用來自 LZ77 的無失真壓縮演算法，壓縮比高，圖片體積小。PNG 支援透明效果，最高可以定義 256 個透明層次，使得影像邊緣能與任何背景平滑地融合，進一步徹底消除鋸齒邊緣。它還原生地為網路傳輸而設計，PNG 圖片會在瀏覽器下載之前提供一個基本的影像內容，隨著圖像資料的下載而逐漸清晰起來。PNG 格式有 8 位元、16 位元和 32 位元三種形式，其中 PNG32 支援 24 位元色彩（約 1600 萬色）和 8 位元透明度，可以再現豐富的真實色彩。

2.2.5 色彩空間

一幅數位影像無論以什麼格式儲存，其顏色的豐富程度決定了它表現和還原真實世界的能力。人們為方便描述和操作各種顏色，從色彩的不同組成方法出發，研究出了很多色彩模型，也就是這裡所說的色彩空間。色彩空間從提出至今已有上百種之多，平時我們常用到的色彩空間包含 RGB、CMYK、YUV、YIQ、HSB/HSV/HSL 和 CIE 等。

RGB 色彩空間源於三原色理論，任何顏色都可以透過紅色（Red）、綠色（Green）、藍色（Blue）3 種原色相加混色而成。RGB 這 3 個分量的數值以亮度來表示，通常它們各有 256（0～255）級亮度。當 3 個分量的亮度最暗時，即 RGB（0, 0, 0）表示黑色；當 3 個分量的亮度最強時，即 RGB（255, 255, 255）表示白色；當 R 分量亮度最大，而其他分量沒有亮度時，即 RGB（255, 0, 0）表示紅色；當 G 分量亮度最大，而其他分量沒有亮度時，即 RGB（0, 255, 0）表示綠色；當 B 分量亮度最大，而其他分量沒有亮度時，即 RGB（0, 0, 255）表示藍色。

RGB 色彩空間是以物體發光定義為基礎的，廣泛地運用於顯示裝置，它更易於被硬體裝置所採納，但它的原理與人的視覺感受並不太相同。為此，人們又提出了 HSB/HSV/HSL/HSI 色彩空間。H（Hue）代表色調，表示人的視覺系統對不同顏色的感受，如綠色、黑色，也可表示一定的顏色範圍，如暖色調、冷色調，決定了顏色的基本特徵。S（Saturation）代表飽和度，表示顏色的純度，即摻入白光的程度，飽和度越大，顏色越鮮豔。B（Brightness）、V（Value）、L（Lightness）、I（Intensity）都是對亮度、明度的度量，表現光的強和弱。

CMYK 色彩空間主要用於出版印刷業，它還是利用三色混色的原理，使用 C（Cyan，青色）、M（Magenta，品紅色）、Y（Yellow，黃色）加上

K（Black，黑色），4 種顏色混色覆蓋而形成豐富的色彩。CMYK 這 4 個分量的數值代表油墨的濃淡，不同於 RGB 是一種依靠反光的色彩模式。

YUV、YIQ、YCbCr 和 YPbPr 色彩空間用於電視系統。Y 均表示亮度，為了減小訊號所佔用的頻寬，其他兩個分量採用 RGB 分量與 Y 的色差表示。YUV 是歐洲電視系統所採用的一種顏色表示法，是 PAL和 SECAM 模擬彩色電視制式採用的色彩空間。其中 U 是 B 和 Y 的一種差值計算，V 是 R 與 Y 的一種差值計算。YIQ 是 NTSC（National Television Standard Committee）電視系統標準，被北美電視系統所採用。YIQ 比 YUV 的差值計算方式更優，所佔頻寬更低。YCbCr 將 YUV的演算法進行了改進，是對 YUV 進行 Gamma 修正的結果。YCbCr 在電腦領域應用很廣，JPEG 和 MPEG 均採用它作為色彩空間，普遍地應用於 DVD、數位電視、攝影機和數位相機中。YCbCr 經過縮放和偏移計算後產生了 YPbPr。

1931 年 9 月，在國際照明委員會（Commission Internationale de l'Eclairage，CIE）的一次會議上，研究人員在 RGB 模型基礎之上，利用數學計算從真實基色推導出理論上的三原色，並提出了一種新的色彩空間CIEXYZ，使染料、印刷等工業能夠明確地指定顏色。在後續的會議上又不斷對其存在的問題進行修改，產生了 CIELUV 和 CIELAB。

2.2.6 影像基本操作

1 取得 JDK 原生支援影像格式

由於智慧財產權等原因，每個版本的 JDK 原生支援的影像格式並不太一樣。作者使用的 JDK1.8 對 JPG、BMP、GIF、PNG、WBMP 和 JPEG格式的影像提供原生的讀寫支援。如果要對其他格式的影像進行讀寫，需要使用 JAI 或協力廠商函數庫。如何取得 JDK 原生支援影像格式，見程式 2-1。

⚊ 程式 2-1

```java
public void getReadWriteFormat() {
    // 讀取取檔案格式副檔名
  String[] readSuffixes = ImageIO.getReaderFileSuffixes();
    // 寫入檔案格式副檔名
  String[] writeSuffixes = ImageIO.getWriterFileSuffixes();
    String canReadFormat = "";
    String canWriteFormat = "";
    for (int i = 0; i < readSuffixes.length; i++) {
        canReadFormat += readSuffixes[i] + ",";
    }
    for (int j = 0; j < writeSuffixes.length; j++) {
        canWriteFormat += writeSuffixes[j] + ",";
    }
    // 去除最後一個逗點
  canReadFormat = canReadFormat.substring(0, canReadFormat.length() - 1);
    canWriteFormat = canWriteFormat.substring(0,
            canWriteFormat.length() - 1);
    System.out.println("JDK 支援對 " + canReadFormat + " 格式讀取 ");
    System.out.println("JDK 支援對 " + canWriteFormat + " 格式寫入 ");
}
```

2 取得影像資訊

javax.imageio.ImageIO 類別 read 方法傳回了一個 BufferedImage 物件，該物件代表已經載入記憶體緩衝區的影像檔。BufferedImage 提供了一系列方法取得影像寬、高，色彩空間類型及分量數、像素位數、透明度等資訊。為解釋實際色彩空間、透明度的意義，提供了 String

getColorSpaceName(int type)、String getTransparencyName(int type)、
String getImageTypeName (int type) 函數，見程式 2-2、程式 2-3 和程式
2-4，取得影像資訊見程式 2-5。

⌛ 程式 2-2

```java
private String getColorSpaceName(int type) {
    String name = "";
    switch (type) {
        case 0:
            name = "TYPE_XYZ，XYZ 色彩空間的任意顏色系列。";
            break;
        case 1:
            name = "TYPE_Lab，Lab 色彩空間的任意顏色系列。";
            break;
        case 2:
            name = "TYPE_Luv，Luv 色彩空間的任意顏色系列。";
            break;
        case 3:
            name = "TYPE_YCbCr，YCbCr 色彩空間的任意顏色系列。";
            break;
        case 4:
            name = "TYPE_Yxy，Yxy 色彩空間的任意顏色系列。";
            break;
        case 5:
            name = "TYPE_RGB，RGB 色彩空間的任意顏色系列。";
            break;
        case 6:
            name = "TYPE_GRAY，GRAY 色彩空間的任意顏色系列。";
            break;
        case 7:
            name = "TYPE_HSV，HSV 色彩空間的任意顏色系列。";
            break;
```

```
case 8:
    name = "TYPE_HLS,HLS 色彩空間的任意顏色系列。";
    break;
case 9:
    name = "TYPE_CMYK,CMYK 色彩空間的任意顏色系列。";
    break;
case 11:
    name = "TYPE_CMY,CMY 色彩空間的任意顏色系列。";
    break;
case 12:
    name = "TYPE_2CLR,Generic 2 分量色彩空間。";
    break;
case 13:
    name = "TYPE_3CLR,Generic 3 分量色彩空間。";
    break;
case 14:
    name = "TYPE_4CLR,Generic 4 分量色彩空間。";
    break;
case 15:
    name = "TYPE_5CLR,Generic 5 分量色彩空間。";
    break;
case 16:
    name = "TYPE_6CLR,Generic 6 分量色彩空間。";
    break;
case 17:
    name = "TYPE_7CLR,Generic 7 分量色彩空間。";
    break;
case 18:
    name = "TYPE_8CLR,Generic 8 分量色彩空間。";
    break;
case 19:
    name = "TYPE_9CLR,Generic 9 分量色彩空間。";
```

```
            break;
        case 20:
            name = "TYPE_ACLR，Generic 10 分量色彩空間 ";
            break;
        case 21:
            name = "TYPE_BCLR，Generic 11 分量色彩空間。";
            break;
        case 22:
            name = "TYPE_CCLR，Generic 12 分量色彩空間。";
            break;
        case 23:
            name = "TYPE_DCLR，Generic 13 分量色彩空間。";
            break;
        case 24:
            name = "TYPE_ECLR，Generic 14 分量色彩空間。";
            break;
        case 25:
            name = "TYPE_FCLR，Generic 15 分量色彩空間。";
            break;
    }
    return name;
}
```

⧗ 程式 2-3

```
private String getTransparencyName(int type) {
    String name = "";
    switch (type) {
        case 1:
            name = "OPAQUE，表示保障完全不透明的圖像資料，表示所有像素的 Alpha
值都為 1.0。";
            break;
        case 2:
```

```
              name = "BITMASK，表示保障完全不透明的圖像資料（Alpha 值為 1.0）或
完全透明的影像數
據（Alpha 值為 0.0）。";
              break;
        case 3:
              name = "TRANSLUCENT，表示包含或可能包含位於 0.0 和 1.0（含兩者）
之間的任意 Alpha 值的圖像資料。";
              break;
    }
    return name;
}
```

⧗ 程式 2-4

```
private String getImageTypeName(int type) {
    String name = "";
    switch (type) {
        case 0:
              name = "TYPE_CUSTOM，沒有識別出影像類型，因此它必定是一個自訂影像。";
              break;
        case 1:
              name = "TYPE_INT_RGB，表示一個影像，它具有合成整數像素的 8 位元 RGB
顏色分量。";
              break;
        case 2:
              name = "TYPE_INT_ARGB，表示一個影像，它具有合成整數像素的 8 位元
RGBA 顏色分量。";
              break;
        case 3:
              name = "TYPE_INT_ARGB_PRE，表示一個影像，它具有合成整數像素的 8 位
元 RGBA 顏色分量。";
              break;
        case 4:
```

```
        name = "TYPE_INT_BGR，表示一個具有 8 位元 RGB 顏色分量的影像，對應於
Windows 或 Solaris 風格的 BGR 色彩模型，具有包裝為整數像素的 Blue、Green 和 Red 三
種顏色。";
        break;
    case 5:
        name = "TYPE_3BYTE_BGR，表示一個具有 8 位元 RGB 顏色分量的影像，對應
於 Windows 風格的 BGR 色彩模型，具有用 3 位元組儲存的 Blue、Green 和 Red 三種顏色。";
        break;
    case 6:
        name = "TYPE_4BYTE_ABGR，表示一個具有 8 位元 RGBA 顏色分量的影像，
具有用 3 位元組儲存的 Blue、Green 和 Red 顏色以及 1 位元組的 Alpha。";
        break;
    case 7:
        name = "TYPE_4BYTE_ABGR_PRE，表示一個具有 8 位元 RGBA 顏色分量的影
像，具有用 3 位元組儲存的 Blue、Green 和 Red 顏色以及 1 位元組的 Alpha。";
        break;
    case 8:
        name = "TYPE_USHORT_565_RGB，表示一個具有 5-6-5 RGB 顏色分量
（5 位 red、6 位元 green、5 位 blue）的影像，不帶 Alpha。";
        break;
    case 9:
        name = "TYPE_USHORT_555_RGB，表示一個具有 5-5-5 RGB 顏色分量
（5 位 red、5 位 green、5 位 blue）的影像，不帶 Alpha";
        break;
    case 10:
        name = "TYPE_BYTE_GRAY，表示無號 byte 灰階級影像（無索引）。";
        break;
    case 11:
        name = "TYPE_USHORT_GRAY，表示一個無號 short 灰階級影像（無索引）。";
        break;
    case 12:
        name = "TYPE_BYTE_BINARY，表示一個不透明的以位元組包裝的 1、2 或 4
位元影像。";
```

```
            break;
        case 13:
            name = "TYPE_BYTE_INDEXED，表示帶索引的位元組影像。";
            break;
        }
    return name;
}
```

⏳ 程式 2-5

```
public void readImageInfo(String imageName) throws IOException {
    File file = new File(imageName);
    BufferedImage image = ImageIO.read(file);
    // 影像寬
    int width = image.getWidth();
    // 影像高
    int height = image.getHeight();
    int imageType = image.getType();
    ColorModel colorModel = image.getColorModel();
    // 每一像素位元數
    int bitsPerPixel = colorModel.getPixelSize();
    // 顏色分量數
    int colorNum = colorModel.getNumComponents();
    ColorSpace colorSpace = colorModel.getColorSpace();
    // 色彩空間類型
    int colorSpaceType = colorSpace.getType();
    // 色彩空間分量數
    int colorSpaceNum = colorSpace.getNumComponents();
    // 透明度
    int transparency = colorModel.getTransparency();
    // 支援 Alpha 分量嗎？
    boolean hasAlpha = colorModel.hasAlpha();
    // 預乘 Alpha 分量嗎？
```

```
    boolean isAlphaPre = colorModel.isAlphaPremultiplied();
    System.out.println("影像" + imageName + "資訊如下:");
    System.out.println("影像的寬度為" + width + "像素,高度為" + height +
"像素");
    System.out.println("影像類型:" + getImageTypeName(imageType));
    System.out.println("一個像素用" + bitsPerPixel + "位表示");
    System.out.println("顏色分量數(包含Alpha分量在內,不支援的除外):" +
colorNum);
    System.out.println("色彩空間:" + getColorSpaceName(colorSpaceType));
    System.out.println("色彩空間分量數:" + colorSpaceNum);
    System.out.println("透明度:" + getTransparencyName(transparency));
    System.out.println(supportAlpha(hasAlpha) + "Alpha分量");
    if (hasAlpha) {
        System.out.println("像素值" + preMultiplied(isAlphaPre) + "預乘
Alpha值");
    }
}
```

Alpha 分量用於指定影像的透明度。例如：一個 16 位元的影像，5 位元表示紅色分量，5 位元表示綠色分量，5 位元表示藍色分量，剩餘 1 位元表示 Alpha 分量。這樣一來，Alpha 分量只能是 1 或 0，即不透明、透明兩種狀態。一個 32 位元的影像，每 8 位元分別表示紅、綠、藍和 Alpha 分量，那麼透明度就可以分 256 來表示。

3 操作影像像素值

Raster 類別表示像素矩形陣列。Raster 封裝儲存樣本值的 DataBuffer，以及描述如何在 DataBuffer 中定位指定樣本值的 SampleModel。我們可以透過 Raster 的 getPixel 和 setPixel 方法來讀寫某點的像素值，程式如下。

⧗ 程式 2-6

```java
public void operateImagePixel(String imageName) throws IOException {
    File file = new File(imageName);
    BufferedImage image = ImageIO.read(file);
    WritableRaster raster = image.getRaster();
    int[] pixel = new int[3];
    // 讀取點 (100,100) 處的像素值
    raster.getPixel(100, 100, pixel);
    System.out.println("影像座標 (100,100) 處像素為:" + "RGB(" + pixel[0]
        + "," + pixel[1] + "," + pixel[2] + ")");
    pixel[0] = 0;//R
    pixel[1] = 0;//G
    pixel[2] = 0;//B
    // 把從 (0,0) 到 (99,99) 的範圍設定為黑色
    for (int x = 0; x < 100; x++) {
        for (int y = 0; y < 100; y++) {
            raster.setPixel(x, y, pixel);
        }
    }
    File outFile = new File("resource/set_pixels_image.jpg");
    ImageIO.write(image, "jpg", outFile);
}
```

❹ 轉換影像格式

利用 ImageIO 類別的讀寫功能,可以很便捷地在支援的影像格式間轉換,程式如下。

⧗ 程式 2-7

```java
public void convertImageFormat(String imageName) throws IOException {
    File file = new File(imageName);
    BufferedImage image = ImageIO.read(file);
```

```
    String convertedImageName = "resource/converted_image_name.bmp";
    File convertedFile = new File(convertedImageName);
    ImageIO.write(image, "bmp", convertedFile);
}
```

5 轉換色彩空間

JDK 提供了 ColorConvertOp 類別，用於對原影像中的資料進行逐像素的顏色轉換，並預置了 CS_sRGB、CS_LINEAR_RGB、CS_CIEXYZ、CS_GRAY 和 CS_PYCC 這 5 種色彩空間。我們可以對以上 5 種預置色彩空間進行精確轉換，實作方式見程式 2-8。如果要對這 5 種以外色彩空間進行轉換，就需要自己讀取 ICC 檔案。

▌ 程式 2-8

```
public void convertImageColorSpace(String imageName) throws IOException {
    File file = new File(imageName);
    BufferedImage image = ImageIO.read(file);
    ColorConvertOp colorConvert = new ColorConvertOp(
            ColorSpace.getInstance(ColorSpace.CS_CIEXYZ), null);
    BufferedImage convertedImage = colorConvert.filter(image, null);
    System.out.println(getColorSpaceName(convertedImage.getColorModel()
            .getColorSpace().getType()));
}
```

在 C:/Windows/System32/spool/drivers/color/ 路徑下存在很多 ICC 檔案，如圖 2-4 所示。透過 ICC_Profile 類別的 getInstance（String fileName）方法，載入 ICC 檔案即可獲得 ICC_Profile 物件，然後把它輸入 ICC_ColorSpace（ICC_Profile profile）建置函數，建立出 ColorSpace 物件後，就可以利用 ColorConvertOp 類別進行色彩空間轉換了，見程式 2-9。

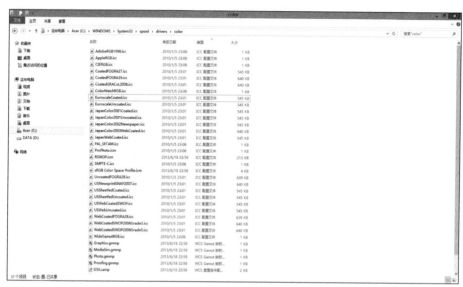

圖 2-4 ICC 檔案

程式 2-9

```java
public void convertImageColorSpaceICC(String imageName, String iccName)
        throws IOException {
    File file = new File(imageName);
    BufferedImage image = ImageIO.read(file);
    String iccPath = "c:/Windows/System32/spool/drivers/color/";
    iccPath += iccName;
    ColorSpace colorSpace = new ICC_ColorSpace(
            ICC_Profile.getInstance(iccPath));
    ColorConvertOp colorConvert = new ColorConvertOp(colorSpace, null);
    BufferedImage convertedImage = colorConvert.filter(image, null);
    System.out.println(getColorSpaceName(convertedImage.getColorModel()
            .getColorSpace().getType()));
}
```

2.3 影像特徵的分類

2.1 節中解釋了人類了解和分辨影像的原理。人腦中記憶了關於事物特徵和名稱的聯繫，所以我們才能一看到某個事物，就立即說出它的名字。同樣，這也是我們不能說出不為我們所知的事物名稱的原因。像顏色、紋理、形狀，這些能讓我們從全域上形容和描述一個事物的特徵叫作全域特徵。

很多人應該都玩過「找不同」的遊戲，如圖 2-5 所示。當我們看到一個事物時，能立即說出它的顏色、形狀和紋理等全域特徵。但像找不同這樣的影像，我們不能立即說出兩個影像的不同點，而需要花時間仔細辨識兩個影像的細節，才能找到它們的不同點。影像在細節或局部上的特徵，我們稱其為局部特徵。

圖 2-5　找不同

如同人類一樣，電腦也是使用顏色、形狀、紋理這種全域特徵，以及用數學方法分析的像 SIFT、SUFT、BRISK 等這種局部特徵來表示影像特徵，並以此來區分它們的，但影像特徵需要具有以下特點。

（1）代表性。影像特徵可以表現這幅影像的特點，能夠代表這幅影像。

（2）穩健性。影像在經過旋轉、縮放、平移等變化後，影像特徵仍能穩
定地代表該影像。

（3）可計算性。影像特徵可以耗費人類能夠接受的時間和資源計算得出。

每種影像特徵分析方法都有一定的優缺點和限制，實際應用中會根據影像的實際特點選擇對應的特徵分析方法，或是選擇幾種影像特徵綜合使用，來加強影像檢索和符合的準確度。

2.4 節和 2.5 節中將對各種典型的影像特徵說明，並透過實作方式程式來逐一說明[1]。

2.4 全域特徵

2.4.1 顏色特徵

顏色特徵分析比較簡單，無須進行大量複雜計算，是一種低複雜度的特徵分析方法。

1 顏色長條圖

顏色長條圖是用於表示影像中各種顏色分佈的一種統計圖，它反映了影像中的顏色種類和這些顏色出現的次數。1991 年，Swain 和 Ballard 最先提出使用顏色長條圖作為影像特徵分析方法，並透過實驗證實將影像

1　本章程式由作者自主撰寫的程式以及部分經過更正和最佳化的開原始程式碼共同組成。

進行旋轉、縮放、模糊等轉換後對顏色長條圖的改變很微小[2]。顏色長條圖性質穩定、計算簡便，但由於它並沒有表現像素的位置特性，常常使幾幅不同的影像卻對應相同或相近的顏色長條圖。以上特點使其特別適用於難以實現自動分割，以及不需要考慮物體空間位置的影像特徵分析。

怎樣產生顏色長條圖呢？

（1）選擇一個色彩空間，以 64bins-RGB 某一分量的顏色長條圖為例，如圖 2-6 所示，程式 2-10 中採用了最常用的 RGB 色彩空間。

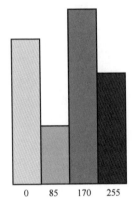

圖 2-6 64bins-RGB 某一分量的顏色長條圖

（2）定義顏色長條圖量化顏色數。計算顏色長條圖，首先需要將色彩空間劃分為許多個小的顏色區域，每個區域稱為一個 bin，這個過程稱為顏色量化。顏色量化有很多種方法，我們通常採用均勻量化，也就是把色彩空間平均分為許多份，也可以採用分群或向量量化以及神經網路的方法。bin 的數量越大，顏色量化消耗的資源也越大，我們需要根據實際應用採用適當的量化顏色數值。

2　Swain, Michael J, Ballard, et al. Color indexing[J]. International Journal of Computer Vision, 1991, 7(1):11-32.

（3）顏色長條圖歸一化。為了使不同解析度的影像能夠進行比較，我們
　　　必須把顏色長條圖進行歸一化，例如將每個 bin 中的值對映為 [0, 1]
　　　的範圍。

⧗ **程式 2-10**

```
package com.ai.deepsearch.features.global.color;

import javax.imageio.ImageIO;
import java.awt.*;
import java.awt.image.BufferedImage;
import java.awt.image.WritableRaster;
import java.io.File;
import java.io.IOException;
import java.util.Arrays;

/**
 * 顏色長條圖
 */
public class ColorHistogram {
    // RGB 色彩空間量化為 64 個 bin
    public static int DEFAULT_NUMBER_OF_BINS = 64;

    // 轉為 RGB 色彩空間
    private BufferedImage convertRGB8Image(BufferedImage image) {
        int width = image.getWidth();
        int height = image.getWidth();
        if (image.getType() != BufferedImage.TYPE_INT_RGB) {
            BufferedImage convertedImage = new BufferedImage(width,
height, BufferedImage.TYPE_INT_RGB);
            Graphics graphics = convertedImage.getGraphics();
            graphics.drawImage(image, 0, 0, null);
            graphics.dispose();
```

```java
        return convertedImage;
    }
    return image;
}


//  歸一化
private void normalize(int[] histogram) {
    int max = 0;
    for (int i = 0; i < histogram.length; i++) {
        max = Math.max(histogram[i], max);
    }
    for (int i = 0; i < histogram.length; i++) {
        histogram[i] = (histogram[i] * 255) / max;
    }
}


/***
 * 顏色量化
 * 每個分量 4 個 bin，共 64 個
 */
private int quantize(int[] pixel) {
    int bin = 0;
    bin = (int) Math.round((double) pixel[2] / 85d)
            + (int) Math.round((double) pixel[1] / 85d) * 4
            + (int) Math.round((double) pixel[0] / 85d) * 4 * 4;
    return bin;
}


public String getHistogramRepresentation(int[] histogram) {
    String histogramString = "";
    for (int i = 0; i < histogram.length; i++) {
        if(i==histogram.length-1) {
            histogramString += String.valueOf(histogram[i]);
```

```java
        } else {
            histogramString += String.valueOf(histogram[i])+",";
        }
    }
    return histogramString;
}


public int[] computeCH(String imageName) throws IOException {
    File file = new File(imageName);
    BufferedImage image = ImageIO.read(file);
    image = convertRGB8Image(image);
    int width = image.getWidth();
    int height = image.getHeight();
    int[] histogram = new int[DEFAULT_NUMBER_OF_BINS];
    Arrays.fill(histogram, 0);
    WritableRaster raster = image.getRaster();
    int[] pixel = new int[3];
    for (int x = 0; x < width; x++) {
        for (int y = 0; y < height; y++) {
            raster.getPixel(x, y, pixel);
            histogram[quantize(pixel)]++;
        }
    }
    normalize(histogram);
    return histogram;
}


public static void main(String args[]) {
    try {
        String imageName = "resource/image_name_rgb8.jpg";
        ColorHistogram ch = new ColorHistogram();
        int[] histogram = ch.computeCH(imageName);
        System.out.println("64bins 顏色長條圖特徵值為:"
```

```
                + ch.getHistogramRepresentation(histogram));
        } catch (IOException e) {
            // TODO Auto-generated catch block
            e.printStackTrace();
        }
    }
}
```

2 顏色矩

顏色矩是一種更為簡單，但有效的顏色特徵。Stricker 和 Orengo 在 1995 年依據影像中的任何顏色分佈都可以用矩來表示的數學原理，提出了顏色矩的方法[3]。顏色矩又可分為一階矩（平均數）、二階矩（方差）和三階矩（偏度），其數學運算式如式（2-1）所示。一幅 RGB 影像有 3 個顏色分量，每個分量上各有 3 個顏色矩，總共 9 個值，實作方式見程式 2-11。

$$\mu = \frac{1}{N}\sum_{j=1}^{N} p_j$$

$$\sigma = \left(\frac{1}{N}\sum_{j=1}^{N}(p_j - \mu)^2\right)^{\frac{1}{2}} \qquad (2\text{-}1)$$

$$s = \left(\frac{1}{N}\sum_{j=1}^{N}(p_j - \mu)^3\right)^{\frac{1}{3}}$$

同顏色長條圖相比，顏色矩計算更為簡便，無須對特徵進行量化。在實際應用中，顏色矩通常和其他特徵結合使用，一般在使用其他影像特徵前造成縮小範圍的作用。

3　Stricker A M A,Orengo M.Similarity of Color Images[J].Proc Spie Storage & Retrieval for Image & Video Databases，1995，2420：381-392.

⧗ 程式 2-11

```java
package com.ai.deepsearch.features.global.color;

import javax.imageio.ImageIO;
import java.awt.image.BufferedImage;
import java.awt.image.WritableRaster;
import java.io.File;
import java.io.IOException;

/**
 * 顏色矩
 */
public class ColorMoments {
    // 計算顏色矩
    public double[][] computeColorMoments(String imageName) throws
IOException {
        File file = new File(imageName);
        BufferedImage image = ImageIO.read(file);
        int width = image.getWidth();
        int height = image.getHeight();
        int pixelsSize = width * height;
        WritableRaster raster = image.getRaster();
        int[] pixel = new int[3];
        int sumR = 0;
        int sumG = 0;
        int sumB = 0;
        for (int x = 0; x < raster.getWidth(); x++) {
            for (int y = 0; y < raster.getHeight(); y++) {
                raster.getPixel(x, y, pixel);
                sumR += pixel[0];
                sumG += pixel[1];
                sumB += pixel[2];
```

```
        }
    }
    // 一階矩
    double meanR = sumR / pixelsSize;
    double meanG = sumG / pixelsSize;
    double meanB = sumB / pixelsSize;
    // 二階矩
    double stdR = 0;
    double stdG = 0;
    double stdB = 0;
    // 三階矩
    double skwR = 0;
    double skwG = 0;
    double skwB = 0;

    for (int x = 0; x < raster.getWidth(); x++) {
        for (int y = 0; y < raster.getHeight(); y++) {
            raster.getPixel(x, y, pixel);
            stdR += Math.pow(pixel[0] - meanR, 2);
            stdG += Math.pow(pixel[1] - meanG, 2);
            stdB += Math.pow(pixel[2] - meanB, 2);

            skwR += Math.pow(pixel[0] - meanR, 3);
            skwG += Math.pow(pixel[1] - meanG, 3);
            skwB += Math.pow(pixel[2] - meanB, 3);
        }
    }

    stdR = Math.sqrt(stdR / pixelsSize);
    stdG = Math.sqrt(stdG / pixelsSize);
    stdB = Math.sqrt(stdB / pixelsSize);
```

```java
        skwR = Math.cbrt(skwR / pixelsSize);
        skwG = Math.cbrt(skwG / pixelsSize);
        skwB = Math.cbrt(skwB / pixelsSize);

        double[][] moments = new double[3][3];
        moments[0][0] = meanR;
        moments[0][1] = meanG;
        moments[0][2] = meanB;

        moments[1][0] = stdR;
        moments[1][1] = stdG;
        moments[1][2] = stdB;

        moments[2][0] = skwR;
        moments[2][1] = skwG;
        moments[2][2] = skwB;

        return moments;
    }


    public static void main(String args[]) {
        String imageName = "resource/image_name_rgb8.jpg";
        ColorMoments colorMoments = new ColorMoments();
        try {
            double[][] moments = colorMoments.computeColorMoments
(imageName);
            System.out.println("顏色矩為一階矩 R:" + moments[0][0] + ",
一階矩 G:"
                    + moments[0][1] + ",一階矩 B:" + moments[0][2] + ",
二階矩 R:"
                    + moments[1][0] + ",二階矩 G:" + moments[1][1] + ",
二階矩 B:"
```

```
                    + moments[1][2] + ",三階矩 R:" + moments[2][0] + ",
三階矩 G:"
                    + moments[2][1] + ",三階矩 B:" + moments[2][2]);
    } catch (IOException e) {
        // TODO Auto-generated catch block
        e.printStackTrace();
    }
}
}
```

3 顏色聚合向量

顏色長條圖、顏色矩等以顏色分佈為基礎的影像特徵分析方式均未考慮像素空間位置資訊，會造成同一顏色分佈會對應多個影像的問題。如圖 2-7 所示，左右兩邊是兩張明顯不同的影像，但它們的顏色長條圖卻是相同的。

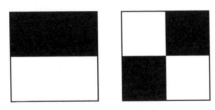

圖 2-7 顏色長條圖相同的兩個影像

為了解決上述問題，Pass 提出了顏色聚合向量（Color Coherence Vector）的概念[4]。顏色聚合向量是顏色長條圖加入色彩空間分佈資訊後的一種演變形式。它將顏色長條圖的每個 bin 中的像素分為兩種 -- 聚合像素和非聚合像素。如果同一 bin 中的像素所佔據的連通區域大於規定

---

4 Pass G, Zabih R, Miller J. Comparing images using color coherence vectors[C]// ACM International Conference on Multimedia. ACM, 1997:65-73.

的設定值，則該區域的像素都屬於聚合像素，否則屬於非聚合像素。假設 bin_i 中的連通像素數目為 α_i，非連通像素數目為 β_i，bin_i 中的像素總數為 $\alpha_i+\beta_i$。一幅影像的顏色聚合向量可表示為 $<(\alpha_1, \beta_1)，(\alpha_2, \beta_2)，\cdots，(\alpha_n, \beta_n)>$。經過進一步思考，我們可以得知 $<(\alpha_1+\beta_1)，(\alpha_2+\beta_2)，\cdots，(\alpha_n+\beta_n)>$ 正是該影像的顏色長條圖。這也從另一個方面證實了顏色聚合向量是由顏色長條圖演變而來的。

建置顏色聚合向量的過程如下（實作方式參考程式 2-12）。

（1）首先將影像統一尺寸。

（2）對影像使用 3×3 範本進行高斯模糊，用 8 鄰域平均值替代原值。

（3）量化色彩空間，減少顏色數目。

（4）依據連通性統計結果將每個像素分為聚合和非聚合。

（5）統計各量化顏色的聚合和非聚合像素數量。

（6）歸一化。

⧗ 程式 2-12

```
package com.ai.deepsearch.features.global.color;

import javax.imageio.ImageIO;
import java.awt. *;
import java.awt.image.BufferedImage;
import java.awt.image.ConvolveOp;
import java.awt.image.Kernel;
import java.io.File;
import java.io.IOException;

/**
 * 顏色聚合向量
```

```java
*/
public class ColorCoherenceVector {
    private int[] currImage;
    private int[][] colorTagged;
    private int differentAreasNum;
    private double[] alpha;
    private double[] beta;

    private final double t = 0.00000001;
    private final int THRESHOLD = 15;

    // 統一尺寸
    private BufferedImage resizeImage(BufferedImage image) {
        int limit = 400;
        int width = image.getWidth();
        int height = image.getHeight();
        if (width < height) {
            width = width * limit / height;
            height = limit;
        } else {
            height = height * limit / width;
            width = limit;
        }
        BufferedImage resizedImage = new BufferedImage(width, height,
                BufferedImage.TYPE_INT_RGB);
        Graphics2D graphics = (Graphics2D) resizedImage.getGraphics();
        graphics.setRenderingHint(RenderingHints.KEY_ANTIALIASING,
                RenderingHints.VALUE_ANTIALIAS_ON);
        graphics.drawImage(image, 0, 0, width, height, null);
        graphics.dispose();
        return resizedImage;
    }
```

```java
    // 產生高斯核心
    private float[] generateGaussianKernel(int radius, float sigma) {
        float center = (float) Math.floor((radius + 1) / 2);
        float[] kernel = new float[radius * radius];
        float sum = 0;
        for (int y = 0; y < radius; y++) {
            for (int x = 0; x < radius; x++) {
                int offset = y * radius + x;
                float distX = x - center;
                float distY = y - center;
                kernel[offset] = (float) ((1 / (2 * Math.PI * sigma *
sigma)) * Math
                        .exp(-(distX * distX + distY * distY)
                            / (2 * (sigma * sigma))));
                sum += kernel[offset];
            }
        }
        // 歸一化
        for (int i = 0; i < kernel.length; i++)
            kernel[i] /= sum;
        return kernel;
    }

    // 採用 3×3範本進行高斯模糊
    private BufferedImage gaussianBlur(BufferedImage image) {
        ConvolveOp gaussian = new ConvolveOp(new Kernel(3, 3,
                generateGaussianKernel(3, 0.8f)),
ConvolveOp.EDGE_NO_OP, null);
        return gaussian.filter(image, null);
    }

    // 重新量化色彩空間
```

```java
private void colorsReduction(BufferedImage image) {
    int width = image.getWidth();
    int height = image.getHeight();
    currImage = image.getRGB(0, 0, width, height, null, 0, width);
    // 192 的二進位是 11000000，相當於將每個顏色分量由 8bit 轉化為 2bit，也就
是將顏色數目減少為 64
    int flag = 192;
    for (int i = 0; i < currImage.length; ++i) {
        int r = (currImage[i] >> 16) & flag;
        int g = (currImage[i] >> 8) & flag;
        int b = currImage[i] & flag;
        currImage[i] = (r << 16) + (g << 8) + b;
    }
}


// 標記連通性
private void tagColor(int width, int height) {
    colorTagged = new int[height][width];
    differentAreasNum = 0;
    for (int row = 0; row < height; row++) {
        for (int col = 0; col < width; col++) {
            int color = currImage[row * width + col];
            if (row > 0) {
                // 左上角
                if (col > 0) {
                    if (currImage[(row - 1) * width + col - 1] ==
color) {
                        colorTagged[row][col] = colorTagged[row -
1][col - 1];
                        continue;
                    }
                }
                // 上
```

```
                    if (currImage[(row - 1) * width + col] == color) {
                        colorTagged[row][col] = colorTagged[row - 1]
[col];

                        continue;
                    }
                    // 右上角
                    if (col < width - 1) {
                        if (currImage[(row - 1) * width + col + 1] ==
color) {

                            colorTagged[row][col] = colorTagged[row -
1][col + 1];

                            continue;
                        }
                    }
                }
                // 左
                if (col > 0) {
                    if (currImage[row * width + col - 1] == color) {
                        colorTagged[row][col] = colorTagged[row][col - 1];
                        continue;
                    }
                }
                colorTagged[row][col] = differentAreasNum;
                differentAreasNum++;
            }
        }
    }

    private void computeCoherence(int width, int height) {
        int[] count = new int[differentAreasNum];
        int[] color = new int[differentAreasNum];

        for (int x = 0; x < height; x++) {
```

```
        for (int y = 0; y < width; y++) {
            count[colorTagged[x][y]]++;
            color[colorTagged[x][y]] = currImage[x * width + y];
        }
    }


    alpha = new double[64];
    beta = new double[64];


    for (int i = 0; i < differentAreasNum; ++i) {
        // d使用中色彩，代表 24bits RGB
        int d = color[i];
        // 轉換 d 至 6bits RGB, 範圍 0-63
        color[i] = (((d >> 22) & 3) << 4) + (((d >> 14) & 3) << 2)
                + ((d >> 6) & 3);
        if (count[i] < t * width * height || count[i] < THRESHOLD) {
            beta[color[i]]++;
        } else {
            alpha[color[i]]++;
        }
    }
}


// 歸一化
private void normalize(int width, int height) {
    for (int i = 0; i < alpha.length; i++) {
        if (alpha[i] == 0 && beta[i] == 0)
            continue;
        alpha[i] /= width * height;
        beta[i] /= width * height;

    }
}
```

```java
public void computeCCV(String imageName) throws IOException {
    File file = new File(imageName);
    BufferedImage image = ImageIO.read(file);
    image = resizeImage(image);
    image = gaussianBlur(image);
    colorsReduction(image);
    int width = image.getWidth();
    int height = image.getHeight();
    tagColor(width, height);
    computeCoherence(width, height);
    normalize(width, height);
}

public String getCCVRepresentation() {
    String ccv = "";
    for (int i = 0; i < alpha.length; ++i) {
        if (alpha[i] == 0 && beta[i] == 0)
            continue;
        ccv += String.format("%2d (%3f, %3f)%n", i, alpha[i],
beta[i]);
    }
    return ccv;
}

public static void main(String args[]) {
    try {
        String imageName = "resource/image_name_rgb8.jpg";
        ColorCoherenceVector ccv = new ColorCoherenceVector();
        ccv.computeCCV(imageName);
        String ccvString = ccv.getCCVRepresentation();
        System.out.println(ccvString);
    } catch (IOException e) {
        // TODO Auto-generated catch block
```

```
            e.printStackTrace();
        }
    }
}
```

2.4.2 紋理特徵

紋理是一種反映物體表面結構變化的屬性，也就是我們通常所說的「花紋」，例如木紋、雲彩等，如圖 2-8 所示。

圖 2-8 木紋和雲彩的紋理

影像紋理特徵按照分析方法的不同，可分為以幾何為基礎的方法、以結構為基礎的方法、以模型為基礎的方法、以統計為基礎的方法和以訊號處理為基礎的方法 5 種。

以幾何為基礎的方法依賴統計幾何特徵來描述影像紋理，但該方法受到各種限制，人們對它的研究較少。

以結構為基礎的方法，會假設影像是由一系列的紋理基元按照一定的規律或重複性關係組合而成的。這種方法比較適合人工合成紋理，不適合自然紋理。

以模型為基礎的方法，會假設影像紋理是由一定參數控制的分佈模型形成的。統計法和訊號處理法是目前常用的兩種紋理特徵分析方法。統計法歸納影像紋理區域中的某些統計特性，訊號處理法利用傅立葉轉換、Gabor 轉換、小波轉換等方法，將時域訊號轉為頻域訊號後再進行特徵分析。影像搜索中常用到的紋理特徵分析方法，主要有 Tamura 紋理特徵、灰階共生矩陣和 Gabor 紋理特徵等。

1 Tamura 紋理特徵

1978 年，Tamura 等人根據人類對紋理的視覺感知心理學研究，提出了包含粗糙度、比較度、方向度、線像度、規整度和粗略度 6 種屬性的紋理特徵描述法[5]。其中，前 3 個屬性對於影像相似度的區分特別重要。

3 個分量的計算，首先需要將影像灰階化。

粗糙度代表影像紋理粗糙的程度，其計算過程分為以下幾步驟，對應程式 2-13。

（1）計算以影像中的一點 (x, y) 為中心，大小為 $2^k \times 2^k$ 的使用中視窗內像素的平均灰階值，記為 $A_k(x, y)$。在式（2-2）中，$k=0,1,\cdots,5$，$g(i, j)$ 代表使用中視窗內的點 (i, j) 的像素灰階值。

$$A_k(x,y) = \sum_{i=x-2^{k-1}}^{x+2^{k-1}-1} \sum_{j=y-2^{k-1}}^{y+2^{k-1}-1} g(i,j) / 2^{2k} \tag{2-2}$$

（2）在影像內移動使用中視窗，分別計算水平方向和垂直方向上互不重疊的視窗間的平均灰階差。在式（2-3）中，$A_k(x+2^{k-1}, y)$ 表示以 $(x+2^{k-1}, y)$ 為中心，$2^k \times 2^k$ 大小的視窗內的平均灰階值。

---

5 Tamura H, Mori S, Yamawaki T. Textural Features Corresponding to Visual Perception[J]. IEEE Trans.syst.man.cybernet, 1978, 8(6):460-473.

$$E_{k,h}(x,y) = \left| A_k(x + 2^{k-1}, y) - A_k(x - 2^{k-1}, y) \right| \quad \text{水平方向平均灰度差}$$

$$E_{k,v}(x,y) = \left| A_k(x, y + 2^{k-1}) - A_k(x, y - 2^{k-1}) \right| \quad \text{垂直方向平均灰度差}$$

(2-3)

（3）對於點 (x,y)，無論是式（2-3）中的水平方向平均灰階差 $E_{k,h}$，還是垂直方向平均灰階差 $E_{k,v}$，都要找出能使 E 最大的 k 值來設定最佳尺寸 $S_{\text{best}}(x, y) = 2^k$。

（4）計算影像中每個像素點的 S_{best} 值，累加後求平均值就獲得影像的粗糙度 F_{crs}。

$$F_{crs} = \frac{1}{m \times n} \sum_{i=1}^{m} \sum_{j=1}^{n} S_{\text{best}}(i, j)$$

(2-4)

下面來計算比較度，影像比較度是統計像素平均值、方差、峰態情況而得的。在式（2-5）中，σ 表示像素灰階值的標準差，α_4 表示灰階值的峰態，它描述了資料在中心的聚集程度。

$$F_{con} = \frac{\sigma}{\alpha_4^{1/4}}$$

$$\sigma = \sqrt{\frac{1}{n} \sum_{i=1}^{n} (g_i - \overline{g}_i)^2}$$

$$\alpha_4 = \frac{\mu_4}{\sigma^4}, \ \mu_4 = \frac{1}{n} \sum_{i=1}^{n} (g_i - \overline{g}_i)^4$$

(2-5)

方向度的計算過程如下。

（1）首先計算每個像素點的梯度 ΔG 和梯度方向 θ：

$$|\Delta G| = (|\Delta H| + |\Delta V|)/2$$

$$\theta = \arctan\left(\frac{\Delta V}{\Delta H}\right) + \pi/2$$

(2-6)

在式（2-6）中，ΔH 和 ΔV 分別是影像與下列 3×3 濾波運算元進行旋積的結果。

1	0	−1
1	0	−1
1	0	−1

計算 ΔH 的濾波運算元

1	1	1
0	0	0
−1	−1	−1

計算 ΔV 的濾波運算元

（2）我們把 $0 \sim \pi$ 區域劃分成 n 等距（n 一般設定值 16）。統計當 $|\Delta G| \geq t$（t 一般設定值 12）時，對應的 θ 所在區間像素數量而形成的長條圖 H_D。

$$H_D(k) = \frac{N_{\hat{e}}(k)}{\sum_{i=0}^{n-1} N_{\hat{e}}(k)}, k = 0,1,\cdots,n-1 \tag{2-7}$$

（3）透過計算長條圖 H_D 中峰值的尖銳程度來表示影像紋理整體的方向性，如圖 2-9 所示。方向度計算如式（2-8）所示，其中 n_p 是長條圖 H_D 中峰值的數量，ϕ_p 是第 p 個峰值，ω_p 是包含第 p 個峰值谷之間的範圍，r 是和 ϕ 的量化相關的歸一化係數。

$$F_{dir} = 1 - r \times n_p \times \sum_{p}^{n_p} \sum_{\varphi \in \omega_p} (\varphi - \varphi_p)^2 \times H_D(\varphi) \tag{2-8}$$

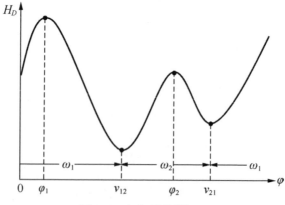

圖 2-9　方向長條圖 H_D

⧗ 程式 2-13

```java
package com.ai.deepsearch.features.global.texture;

import com.ai.deepsearch.utils.ImageUtil;

import javax.imageio.ImageIO;
import java.awt. *;
import java.awt.color.ColorSpace;
import java.awt.image.BufferedImage;
import java.awt.image.ColorConvertOp;
import java.awt.image.WritableRaster;
import java.io.File;
import java.io.IOException;

/**
 * Tamura 特徵
 */
public class Tamura {
    private static final int MAX_IMG_HEIGHT = 64;
    private int histogramBins = 16;
    private double histogramThreshold = 12.0;
    private int[][] grayScales;
    private double[] histogram;
    //Prewitt 梯度運算元
    private static final double[][] prewittFilterH = {{-1, 0, 1}, {-1,
0, 1}, {-1, 0, 1}};
    private static final double[][] prewittFilterV = {{-1, -1, -1}, {0,
0, 0}, {1, 1, 1}};

    /*
     * 計算粗糙度 (Coarseness)
     */
```

```
// 步驟 1. 計算以點 (x,y) 為中心 ，2ᵏ × 2ᵏ 大小的使用中視窗內像素的平均灰階值
private double averageOverNeighborhoods(int x, int y, int k, int
width, int heigh{
    double result = 0;
    double border = Math.pow(2, k);
    int x0 = 0, y0 = 0;

    for (int i = 0; i < border; i++) {
        for (int j = 0; j < border; j++) {
            // x-2ᵏ⁻¹
            x0 = x - (int) Math.pow(2, k - 1) + i;
            // y-2ᵏ⁻¹
            y0 = y - (int) Math.pow(2, k - 1) + j;
            //如果使用中視窗移動到影像以外
            if (x0 < 0) x0 = 0;
            if (y0 < 0) y0 = 0;
            if (x0 >= width) x0 = width - 1;
            if (y0 >= height) y0 = height - 1;

            result = result + grayScales[x0][y0];
        }
    }
    // 2ᵏ× 2ᵏ=2²ᵏ
    result = result / Math.pow(2, 2 * k);
    return result;
}

// 步驟 2. 點 (x,y) 在水平方向上互不重疊的使用中視窗間的平均強度差
private double differencesBetweenNeighborhoodsHorizontal(int x,
int y, int k, int width, int height) {
    double result = 0;
    // |(x+2ᵏ1,y)-(x-2ᵏ1,y)|
    result = Math.abs(this.averageOverNeighborhoods(
```

```
                x + (int) Math.pow(2, k - 1), y, k, width, height)
                - this.averageOverNeighborhoods(x - (int) Math.pow(2, k
- 1),
                y, k, width, height));
        return result;
    }


    // 接步驟 2. 點 (x,y) 在垂直方向上互不重疊的使用中視窗間的平均強度差
    private double differencesBetweenNeighborhoodsVertical(int x, int
y, int k, int width, int height) {
        double result = 0;
        // |(x,y+2ᵏ1)-(x,y-2ᵏ1)|
        result = Math.abs(this.averageOverNeighborhoods(x,
                y + (int) Math.pow(2, k - 1), k, width, height)
                - this.averageOverNeighborhoods(x,
                y - (int) Math.pow(2, k - 1), k, width, height));
        return result;
    }


    // 步驟 3. 點 (x,y)，找出能使平均強度差 ( 無論水平還是垂直方向 ) 最大的 k 值
    private int findBestK(int x, int y, int width, int height) {
        double result = 0, best=0;
        int maxK = 1;

        for (int k = 0; k < 3; k++) {
            best = Math.max(
                    this.differencesBetweenNeighborhoodsHorizontal(x,
y, k, width, height),
                    this.differencesBetweenNeighborhoodsVertical(x, y,
k, width, height));
            if (result < best) {
                maxK = k;
                result = best;
```

```
                }
            }
        return maxK;
    }


    // 粗糙度
    // m 為影像寬度 , n 為影像高度
    private double coarseness(int m, int n) {
        double result = 0;
        for (int i = 1; i < m - 1; i++) {
            for (int j = 1; j < n - 1; j++) {
                result = result + Math.pow(2, this.findBestK(i, j, m,
n));
            }
        }
        result = result / (m * n);
        return result;
    }


    /*
     * 計算比較度 (Constrast)
     */


    // 影像灰階平均值
    private double calculateMean(int width, int height) {
        double mean = 0;

        for (int x = 0; x < width; x++) {
            for (int y = 0; y < height; y++) {
                mean = mean + this.grayScales[x][y];
            }
        }
        mean = mean / (width * height);
```

```java
        return mean;
    }


    // 影像灰階標準差
    private double calculateSigma(double mean, int width, int height) {
        double result = 0;

        for (int x = 0; x < width; x++) {
            for (int y = 0; y < height; y++) {
                result = result + Math.pow(this.grayScales[x][y] -
mean, 2);
            }
        }
        result = result / (width * height);
        return Math.sqrt(result);
    }


    // 峰態
    private double calculateKurtosis(int width, int height) {
        double alpha4 = 0;
        double mu4 = 0;
        double mean = this.calculateMean(width, height);
        double sigma = this.calculateSigma(mean, width, height);

        for (int x = 0; x < width; x++) {
            for (int y = 0; y < height; y++) {
                mu4 = mu4 + Math.pow(this.grayScales[x][y] - mean, 4);
            }
        }
        mu4 = mu4 / (width * height);
        alpha4 = mu4 / (Math.pow(sigma, 4));
        return alpha4;
    }
```

```java
    // 比較度
    private double contrast(int width, int height) {
        double result = 0;
        double mean = this.calculateMean(width, height);
        double sigma = this.calculateSigma(mean, width, height);

        if (sigma <= 0) return 0;

        double alpha4 = this.calculateKurtosis(width, height);
        result = sigma / (Math.pow(alpha4, 0.25));
        return result;
    }


    /*
     * 計算方向度 (Directionality)
     */

    // 點 (x,y) 與 prewittFilterH 進行旋積
    private double calculateDeltaH(int x, int y) {
        double result = 0;

        for (int i = 0; i < 3; i++) {
            for (int j = 0; j < 3; j++) {
                result += this.grayScales[x - 1 + i][y - 1 + j] *
prewittFilterH[i][j];
            }
        }

        return result;
    }


    // 點 (x,y) 與 prewittFilterV 進行旋積
    private double calculateDeltaV(int x, int y) {
```

```
        double result = 0;

        for (int i = 0; i < 3; i++) {
            for (int j = 0; j < 3; j++) {
                result += this.grayScales[x - 1 + i][y - 1 + j] *
prewittFilterV[i][j];
            }
        }
        return result;
    }

    private int[] directionalHistogram(int width, int height, int
totalBinCount) {
        totalBinCount = 0;
        int binWindow = 0;
        int bin[] = new int[this.histogramBins];

        for (int x = 1; x < width - 1; x++) {
            for (int y = 1; y < height - 1; y++) {
                double deltaV = this.calculateDeltaV(x, y);
                double deltaH = this.calculateDeltaH(x, y);
                double deltaG = (Math.abs(deltaH) + Math.abs(deltaV)) / 2;
                double theta = Math.atan2(deltaV, deltaH);
                //將atan控制在-PI/2至PI/2間
                if (theta < -Math.PI / 2.0) {
                    theta = theta + Math.PI;
                } else if (theta > Math.PI / 2.0) {
                    theta = theta - Math.PI;
                }
                //將theta範圍控制在0至PI間
                theta = theta + Math.PI / 2.0;
                if (deltaG >= this.histogramThreshold) {
                    totalBinCount++;
```

```java
                        // 把 0 至 PI 的區域劃分為 16 等份
                        binWindow = (int) (theta / (Math.PI / this.
histogramBins));

                        if (binWindow == histogramBins) {
                            binWindow = 0;
                        }
                        bin[binWindow]++;
                }
            }
        }
        return bin;
    }


    // 歸一化
    private double[] normalizeHistogram(int[] bin, int totalBinCount) {
        double histogramNormalization[] = new double[histogramBins];
        for (int i = 0; i < histogramBins; i++) {
            histogramNormalization[i] = (double) bin[i] / (double)
totalBinCount;
        }
        return histogramNormalization;
    }


    // 方向度
    private double directionality(int width, int height) {
        int totalBinCount = 0;
        int bin[] = new int[this.histogramBins];
        bin = directionalHistogram(width, height, totalBinCount);
        // 歸一化後的長條圖
        double Hd[] = new double[this.histogramBins];
        Hd = normalizeHistogram(bin, totalBinCount);
        int lastPeak = -1;
        int lastValley = -1;
```

```java
    int peakCount = 0;
    double directionality = 0d;
    for (int i = 0; i < Hd.length - 1; i++) {
        double phiDiff = Hd[i + 1] - Hd[i];
        if (phiDiff > 0 && lastValley == -1) {
            lastValley = i;
            lastPeak = -1;
        } else if (phiDiff < 0 && lastPeak == -1) {
            peakCount++;
            lastPeak = i;
            if (lastValley != -1) {
                for (int j = lastValley; j < i; j++) {
                    directionality += Math.pow(j - i, 2) * Hd[j];
                }
                lastValley = -1;
            }
        } else if (phiDiff < 0) {
            directionality += Math.pow(i - lastPeak, 2) * Hd[i];
        }
    }
    //方向度計算公式
    double r = this.histogramBins / Math.PI;
    directionality = 1 - r * peakCount * directionality;
    return directionality;
}


public String getTamruaRepresentation() {
    StringBuilder sb = new StringBuilder(histogram.length);
    for (int i = 0; i < histogram.length; i++) {
        if(i==histogram.length-1) {
            sb.append(histogram[i]);
        } else {
            sb.append(histogram[i]+",");
```

```java
            }
        }
        return sb.toString().trim();
    }

    public void computeTamrua(String imageName) throws IOException {
        File file = new File(imageName);
        BufferedImage image = ImageIO.read(file);
        // tamura 長條圖
        histogram = new double[3];
        // 轉為灰階色彩空間
        ColorConvertOp colorConvertOp = new ColorConvertOp(image
                .getColorModel().getColorSpace(),
                ColorSpace.getInstance(ColorSpace.CS_GRAY), new
RenderingHints(
                RenderingHints.KEY_COLOR_RENDERING,
                RenderingHints.VALUE_COLOR_RENDER_QUALITY));
        BufferedImage grayImage = colorConvertOp.filter(image, null);
        // 統一影像大小
        grayImage = ImageUtil.scaleImage(grayImage, MAX_IMG_HEIGHT);
        WritableRaster raster = grayImage.getRaster();
        int width = raster.getWidth();
        int height = raster.getHeight();
        int[] pixel = new int[3];
        this.grayScales = new int[width][height];
        for (int x = 0; x < width; x++) {
            for (int y = 0; y < height; y++) {
                raster.getPixel(x, y, pixel);
                this.grayScales[x][y] = pixel[0];
            }
        }

        int grayWidth = grayImage.getWidth();
```

```java
    int grayHeight = grayImage.getHeight();
    //粗糙度
    histogram[0] = this.coarseness(grayWidth, grayHeight);
    //比較度
    histogram[1] = this.contrast(grayWidth, grayHeight);
    //方向度
    histogram[2] = this.directionality(grayWidth, grayHeight);
}


public static void main(String args[]) {
    try {
        String imageName = "resource/image_name_rgb8.jpg";
        Tamura tamura = new Tamura();
        tamura.computeTamrua(imageName);
        System.out.println(tamura.getTamruaRepresentation());
    } catch (IOException e) {
        // TODO Auto-generated catch block
        e.printStackTrace();
    }
}
}
```

2 灰階共生矩陣（Gray-Level Cooccurrence Matrix，GLCM）

1973 年，Haralick 等人在一篇名為 *Textural Features for Image Classification* 的論文中提出了灰階共生矩陣的概念[6]。灰階共生矩陣在原文中被稱為 Gray-Tone Spatial-Dependence Matrices。它描述了紋理灰階值空間分佈的相關性，是影像灰階在一定限制條件下共同出現的機率分佈情況。

---

6 Haralick R M, Shanmugam K, Dinstein I. Textural Features for Image Classification[J]. Systems Man & Cybernetics IEEE Transactions on, 1973, smc-3(6):610-621.

如圖 2-10 所示，一幅影像使用灰階級 0 ～ 3 可表示為矩陣 g。影像 a 指出了座標移動的 4 個方向 -- 水平、主對角線、垂直、副對角線（相對方向計入同一個方向），分別用 0°、45°、90°、135° 來表示，移動的距離為 1 個單位（$d=1$）。$g_1 \sim g_4$ 表示 4 個方向上的灰階共生矩陣，例如 g_1 表示 g 中水平相鄰的灰階共同出現的機率。紅色橢圓圈出的灰階對（0，0）共同出現了 4 次，在矩陣 g_1（0，0）格中用 4 除以總次數 24 而形成的機率表示。

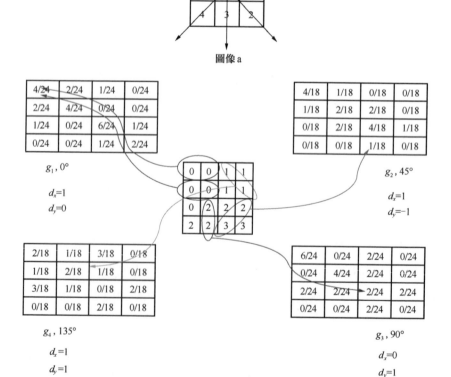

圖 2-10 影像 a 及其灰階共生矩陣

Haralick 從影像的灰階共生矩陣中分析了 28 種特徵值用來代表該影像。這 28 種特徵值有些是強相關的，我們在應用中通常選取它們的子集來代表影像紋理特徵（以下公式中 $P(i, j)$ 代表灰階共生矩陣中的數值，N_g 代表灰階級，灰階值範圍為 $[1..N_g]$），對應程式 2-14。

（1）角二階矩（Angular Second Moment，ASM），也稱為能量，它代表了影像灰階分佈的均勻程度和紋理的粗細程度。

$$ASM = \sum_{i=1}^{N_g} \sum_{j=1}^{N_g} P(i, j)^2 \tag{2-9}$$

（2）比較度（Contrast）反映了影像的清晰度和紋理溝紋深淺的程度。

$$Contrast = \sum_{i=1}^{N_g} \sum_{j=1}^{N_g} (i - j)^2 P(i, j) \tag{2-10}$$

（3）相關性（Correlation）代表了灰階共生矩陣在水平和垂直方向數值的相似程度，當數值比較均勻時，相關性就大，如果數值相差很大，相關性就小。在式（2-11）中，μ_x、μ_y 代表 x 以及 y 方向的平均值，σ_x、σ_y 代表 x 和 y 方向的標準差。

$$Correlation = \sum_{i=1}^{N_g} \sum_{j=1}^{N_g} \frac{(i - \mu_x)(j - \mu_y)P(i, j)}{\sigma_x \sigma_y} \tag{2-11}$$

（4）同質性（Homogeneity），也叫逆差距，用來度量紋理局部變化的多少。如果同質性大，說明影像紋理的區域間缺少變化。

$$Homogeneity = \sum_{i=1}^{N_g} \sum_{j=1}^{N_g} \frac{P(i, j)}{1 + |i - j|} \tag{2-12}$$

（5）熵（Entropy），灰階共生矩陣的熵是一種廣義上的熵，表示攜帶的
資訊量。如果灰階共生矩陣中的像素點對（i, j）出現的機率一樣，
那麼它們攜帶的資訊量是一致的，這時熵最大。

$$Entropy = -\sum_{i=1}^{N_g}\sum_{j=1}^{N_g} P(i,j)\ln(P(i,j))$$ (2-13)

▌ 程式 2-14

```java
package com.ai.deepsearch.features.global.texture;

import com.ai.deepsearch.utils.ImageUtil;

import javax.imageio.ImageIO;
import java.awt. *;
import java.awt.color.ColorSpace;
import java.awt.image.BufferedImage;
import java.awt.image.ColorConvertOp;
import java.awt.image.WritableRaster;
import java.io.File;
import java.io.IOException;

/**
 * 灰階共生矩陣
 */
public class GLCM {
    private static final int MAX_IMG_HEIGHT = 64;
    private static final int DEAFAULT_GRAY_LEVEL = 10;
    // 灰階共生矩陣特徵值
    private double contrast = 0d;
    private double correlation = 0d;
    private double energy = 0d;
    private double homogeneity = 0d;
```

```java
private double entropy = 0d;

// 計算比較度
private double computeContrast(double[][] GLCM) {
    double contrast = 0d;
    int levels = GLCM.length;
    for (int x = 0; x < levels; x++) {
        for (int y = 0; y < levels; y++) {
            contrast += (x - y) * (x - y) * GLCM[x][y];
        }
    }
    return contrast;
}

// 計算相關性
private double computeCorrelation(double[][] GLCM) {
    double correlation = 0d;
    int levels = GLCM.length;

    double meanX = 0;
    double meanY = 0;
    for (int x = 0; x < levels; x++) {
        for (int y = 0; y < levels; y++) {
            meanX += x * GLCM[x][y];
            meanY += y * GLCM[x][y];
        }
    }

    double stdX = 0;
    double stdY = 0;
    for (int x = 0; x < levels; x++) {
        for (int y = 0; y < levels; y++) {
            stdX += (x - meanX) * (x - meanX) * GLCM[x][y];
```

```java
                    stdY += (y - meanY) * (y - meanY) * GLCM[x][y];
            }
        }
        stdX = Math.sqrt(stdX);
        stdY = Math.sqrt(stdY);

        for (int x = 0; x < levels; x++) {
            for (int y = 0; y < levels; y++) {
                double num = (x - meanX) * (y - meanY) * GLCM[x][y];
                double denum = stdX * stdY;
                correlation += num / denum;
            }
        }
        return correlation;
    }

    // 計算能量
    private double computeEnergy(double[][] GLCM) {
        double energy = 0d;
        int levels = GLCM.length;
        for (int x = 0; x < levels; x++) {
            for (int y = 0; y < levels; y++) {
                energy += GLCM[x][y] * GLCM[x][y];
            }
        }
        return energy;
    }

    // 計算同質性
    private double computeHomogeneity(double[][] GLCM) {
        double homogeneity = 0d;
        int levels = GLCM.length;
        for (int x = 0; x < levels; x++) {
```

```java
        for (int y = 0; y < levels; y++) {
            homogeneity += GLCM[x][y] / (1 + Math.abs(x - y));
        }
    }
    return homogeneity;
}


// 計算熵
private double computeEntropy(double[][] GLCM) {
    double entropy = 0d;
    int levels = GLCM.length;
    for (int x = 0; x < levels; x++) {
        for (int y = 0; y < levels; y++) {
            if (GLCM[x][y] != 0) {
                entropy += -GLCM[x][y] * Math.log(GLCM[x][y]);
            }
        }
    }
    return entropy;
}

// 將影像灰階值用灰階級數來表示
private static int[][] computeLeveledMatrix(int[][] matrix,
                    int levels,
                    double minLevel, double maxLevel) {
    int[][] leveledMatrix = new int[matrix.length][matrix[0].length];

    for (int x = 0; x < matrix.length; x++) {
        for (int y = 0; y < matrix[0].length; y++) {

            int grayLevel = (int) (Math.floor((matrix[x][y] -
                        minLevel) * levels / maxLevel));
            if (grayLevel < 0) {
```

```
                    grayLevel = 0;
                } else if (grayLevel >= levels) {
                    grayLevel = levels - 1;
                }
                leveledMatrix[x][y] = grayLevel;
            }
        }
        return leveledMatrix;
    }


    // 使用 Haralick 方法產生灰階共生矩陣
    private double[][] computeGLCMHaralick(int[][] leveledMatrix,
                        int degree, int levels) {
        double[][] GLCM = new double[levels][levels];
        int sum = 0;
        int width = leveledMatrix.length;
        int height = leveledMatrix[0].length;
        int dx = 0;
        int dy = 0;
        if (degree == 0) {
            dx = 1;
            dy = 0;
        }
        if (degree == 45) {
            dx = 1;
            dy = -1;
        }
        if (degree == 90) {
            dx = 0;
            dy = 1;
        }
        if (degree == 135) {
            dx = 1;
```

```
            dy = 1;
        }
    for (int x = 0; x < width; x++) {
        for (int y = 0; y < height; y++) {
            // dx,dy 共生點的 x,y 偏移量
            if (x + dx >= 0 && y + dy >= 0 && x + dx < width
                    && y + dy < height) {
                int v1 = leveledMatrix[x][y];
                int v2 = leveledMatrix[x + dx][y + dy];
                // v1 和 v2 相等時,各方向包含相對的兩個方向,例如 0°,包含 0°
                // 方向和 180° 方向,所以記數兩次。
                if (v1 == v2) {
                    sum += 2;
                    GLCM[v1][v2] += 2;
                } else {
                    sum++;
                    GLCM[v1][v2]++;
                }
            }
        }
    }
    // 歸一化
    for (int x = 0; x < levels; x++) {
        for (int y = 0; y < levels; y++) {
            GLCM[x][y] /= sum;
        }
    }
    return GLCM;
}

// 根據灰階共生矩陣計算影像紋理特徵值
public void computeGLCMFeatures(String imageName) throws
IOException {
```

```java
        File file = new File(imageName);
        BufferedImage image = ImageIO.read(file);
        // 轉為灰階色彩空間
        ColorConvertOp colorConvertOp = new ColorConvertOp(image
                .getColorModel().getColorSpace(),
                ColorSpace.getInstance(ColorSpace.CS_GRAY), new
RenderingHints(
                RenderingHints.KEY_COLOR_RENDERING,
                RenderingHints.VALUE_COLOR_RENDER_QUALITY));
        BufferedImage grayImage = colorConvertOp.filter(image, null);
        // 統一影像大小
        grayImage = ImageUtil.scaleImage(grayImage, MAX_IMG_HEIGHT);
        WritableRaster raster = grayImage.getRaster();
        int width = raster.getWidth();
        int height = raster.getHeight();
        int[] pixel = new int[3];
        int[][] grayScales = new int[width][height];
        // 最小、最大像素值
        double minLevel = Double.MAX_VALUE;
        double maxLevel = -1;
        for (int x = 0; x < width; x++) {
            for (int y = 0; y < height; y++) {
                raster.getPixel(x, y, pixel);
                grayScales[x][y] = pixel[0];
                if (pixel[0] > maxLevel)
                    maxLevel = pixel[0];
                if (pixel[0] < minLevel)
                    minLevel = pixel[0];
            }
        }
        int[][] leveledMatrix = computeLeveledMatrix(grayScales,
                DEAFAULT_GRAY_LEVEL, minLevel, maxLevel);
        // 計算 0° 灰階共生矩陣
```

```java
    double[][] GLCM = computeGLCMHaralick(leveledMatrix, 0,
            DEAFAULT_GRAY_LEVEL);
    this.contrast = computeContrast(GLCM);
    this.correlation = computeCorrelation(GLCM);
    this.energy = computeEnergy(GLCM);
    this.homogeneity = computeHomogeneity(GLCM);
    this.entropy = computeEntropy(GLCM);
}


public String getGLCMRepresentation() {
    int featuresLength = 5;
    StringBuilder sb = new StringBuilder(featuresLength);
    sb.append(this.contrast + ",");
    sb.append(this.correlation + ",");
    sb.append(this.energy + ",");
    sb.append(this.homogeneity + ",");
    sb.append(this.entropy);
    return sb.toString().trim();
}


public static void main(String args[]) {
    try {
        String imageName = "resource/image_name_rgb8.jpg";
        GLCM glcm = new GLCM();
        glcm.computeGLCMFeatures(imageName);
        System.out.println(glcm.getGLCMRepresentation());
    } catch (IOException e) {
        // TODO Auto-generated catch block
        e.printStackTrace();
    }
}
}
```

❸ Gabor 小波紋理特徵

法國數學家傅立葉在 19 世紀研究熱傳播時創立了一套數學理論，該理論被後人不斷研究和發展，形成了著名的傅立葉轉換，傅立葉轉換能夠將訊號在時域和頻域間進行相互轉換，基於這一特性，它也作為一種影像特徵分析方法而被廣泛使用。傅立葉轉換只適合處理平穩訊號，根據傅立葉轉換公式 $F(\omega) = \int_{-\infty}^{+\infty} f(t)\mathrm{e}^{-j\omega t}\mathrm{d}t$ 可知，傅立葉轉換只能分析訊號在整個時間域上的頻率特性，無法獲知實際時間的頻率資訊。對於核心醫學、超音波影像等不平穩訊號，它們的頻域特徵是隨時間變化的，傅立葉轉換便無能為力了。正是以傅立葉轉換為基礎的這一缺陷，1946 年英國物理學家 Dennis Gabor 提出了短時傅立葉轉換，又叫視窗傅立葉轉換，如式（2-14）所示。他在此基礎上進一步利用高斯函數作為時間窗對傅立葉轉換進行擴充，提出了 Gabor 轉換，如式（2-15）所示。

$$X(t,\omega) = \int_{-\infty}^{+\infty} x(s)g(s-t)\mathrm{e}^{-j\omega s}\mathrm{d}s \tag{2-14}$$

$g(s)$ 代表窗函數，可以取漢明、高斯等函數。當 $g(s)$ 取高斯函數時，又被稱為 Gabor 轉換。

$$F(t,\omega) = \pi^{-\frac{1}{4}} \int_{-\infty}^{+\infty} f(s)e^{-\frac{(s=t)^2}{2}} e^{-j\omega s}ds \tag{2-15}$$

短時傅立葉轉換的基本思維是把非平穩訊號看作一系列短時平穩訊號的覆蓋，短時性透過在時間軸上可以移動的窗函數設定值實現。可是這一視窗的大小是固定不變的，對於訊號的高頻部分，其波形較窄，時間間隔小，這就需要一個較小的時間視窗來分析訊號；然而在短時傅立葉轉換中，視窗的大小是固定的，使用這個較小的時間視窗來分析波形較寬，時間間隔大的低頻訊號就勉為其難了。經過 Alfred Haar、Paul Levy、Jean Morlet 等科學家的不懈努力，對短時傅立葉轉換進一步改進，逐步形成了小波轉換理論。小波轉換採用適當的母小波，並對母小波進行旋轉、平移、伸縮等轉換，獲得一系列的小波。這一系列的小波可以將訊號轉換到不同的頻率範圍和時間位置，進一步徹底克服了傅立葉轉換以及短時傅立葉轉換的限制。

$$w(s,\tau) = \frac{1}{\sqrt{s}} \int_{-\infty}^{+\infty} x(t)\psi*\left(\frac{t-\tau}{s}\right)\mathrm{d}t \quad (\psi* \text{ 代表母小波複共軛}) \qquad (2\text{-}16)$$

Gabor 小波是將 Gabor 轉換和小波轉換相結合的產物。由於 Gabor 小波具有與人類視覺系統相似的特性，其在電腦視覺、影像特徵分析、模式識別等領域都獲得了廣泛的應用。Gabor 小波特徵分析的基本思維是將 Gabor 函數作為母小波，來實現多尺度、多方向的紋理分析。在實際的應用領域，根據經驗，Gabor 小波通常取不同的參數值。在影像搜索領域，我們參考了 S. Mangijao Singh 的論文 *Comparative study on content based image retrieval based on Gabor texture features at different scales of frequency and orientations* 的參數設定。

一幅大小為 *PXQ* 的影像 *I(x,y)* 的 Gabor 小波轉為：

$$G_{mn}(x,y) = \sum_s \sum_t I(x-s,y-t)\psi_{mn}^*(s,t) \qquad (2\text{-}17)$$

其中，*s* 和 *t* 是高斯核心函數的大小，$\psi*_{mn}$ 是 ψ_{mn} 的複共軛。母小波函數：

$$\psi(x,y) = \frac{1}{2\pi\sigma_x\sigma_y}\exp\left[-\frac{1}{2}\left(\frac{x^2}{\sigma_x^2}+\frac{y^2}{\sigma_y^2}\right)\right]\exp(j2\pi Wx) \qquad (2\text{-}18)$$

經過擴張和旋轉後產生了一系列 Gabor 小波在 ψ_{mn}。在 $\psi_{mn}(x,y) = a^{-m}\psi(\tilde{x},\tilde{y})$ 中，*m* 和 *n* 分別是高斯核心的尺度和方向，*m*=0, 1…, *M*−1，*n*=0, 1…, *N*−1。其中：

$$\tilde{x} = a^{-m}(x\cos\theta + y\sin\theta)$$

$$\tilde{y} = a^{-m}(-x\sin\theta + y\cos\theta)$$

$$\theta = n\pi / N$$

$$a = (U_h / U_l)^{\frac{1}{M-1}}$$

$$W_{m,n} = a^m U_l$$

$$\sigma_{x,m,n} = \frac{(a+1)\sqrt{2\ln 2}}{2\pi a^m (a-1) U_l}$$

$$\sigma_{y,m,n} = \frac{1}{2\pi \tan\left(\dfrac{\pi}{2N}\right)\sqrt{\dfrac{U_h^2}{2\ln 2} - \left(\dfrac{1}{2\pi\sigma_{x,m,n}}\right)^2}} \tag{2-19}$$

$$U_l = 0.05 \quad U_h = 0.4$$

影像 I 經過不同尺度和方向的 Gabor 小波濾波器處理後，可以獲得一組模：

$$E(m,n) = \sum_x \sum_y |G_{mn}(x,y)| \tag{2-20}$$

這些模代表影像紋理在不同方向和尺度下的能量。模的平均數 μ_{mn} 和標準差 σ_{mn} 表示影像紋理特徵的同質性。

$$\mu_{mn} = \frac{E(m,n)}{PXQ}$$

$$\sigma_{mn} = \frac{\sqrt{\sum_x \sum_y \left(|G_{mn}(x,y)| - \mu_{mn}\right)^2}}{PXQ} \tag{2-21}$$

我們使用 μ_{mn} 和 σ_{mn} 組成的陣列表示影像紋理特徵：$f_g = \{\mu_{00}, \sigma_{00}, \mu_{01}, \sigma_{01}, \cdots, \mu_{MN}, \sigma_{MN}\}$。

實作方式如程式 2-15。

⧗ 程式 2-15

```java
package com.ai.deepsearch.features.global.texture;

import com.ai.deepsearch.utils.ImageUtil;

import javax.imageio.ImageIO;
import java.awt.image.BufferedImage;
import java.awt.image.WritableRaster;
import java.io.File;
import java.io.IOException;

/**
 * Gabor 小波
 */
public class Gabor {
    private static final double Uh = 0.4; // 中心頻率上界
    private static final double Ul = 0.05; // 中心頻率下界
    private static final int S = 4, T = 4; // 高斯核心大小
    private static final int M = 5; // 尺度 (scale) 數
    private static final int N = 6; // 方向 (orientation) 數
    private static final int MAX_IMG_HEIGHT = 64;
    // a=(Uh/Ul)^(1/(M-1))
    private static final double a = Math.pow((Uh / Ul), 1.0 / (M - 1));
    private static double[] theta = new double[N];//θ 角為高斯核心旋轉方向
    private static double[] W = new double[M]; // modulation frequency
    private static double[] sigmaX = new double[M]; // 高斯核心 x 方向尺度
    private static double[] sigmaY = new double[M]; // 高斯核心 y 方向尺度
    private static final double LOG2 = Math.log(2); // ln2
    private double[][][][][] gaborWavelet = null;

    public Gabor() {
        preComputedVariables();
```

```java
        }

        // 公式中用到的變數預先計算
        private void preComputedVariables() {
            // θ =nπ/N
            for (int i = 0; i < N; i++) {
                theta[i] = i * Math.PI / N;
            }

            for (int i = 0; i < M; i++) {
                // Wm,n=a^m Ul
                W[i] = Math.pow(a, i) * Ul;
                // σx,m,n
                sigmaX[i] = (a + 1) * Math.sqrt(2 * LOG2)
                        / (2 * Math.PI * Math.pow(a, i) * (a - 1) * Ul);
                // σy,m,n
                sigmaY[i] = 1 / (2 * Math.PI * Math.tan(Math.PI / (2 * N))
                            * Math.sqrt(Math.pow(Uh, 2) / (2 * LOG2)
                            - Math.pow(1 / (2 * Math.PI * sigmaX[i]), 2)));
            }
        }

        // Gabor 小波轉換公式 Gmn(x,y)=I(x-s)(y-t)φmn(s,t)
        private void computeGaborWavelet(int[][] image,
                                         double[][][][][] childWavelets) {
            this.gaborWavelet = new double[image.length - S][image[0].
length - T][M][N][2];
            for (int m = 0; m < M; m++) {
                for (int n = 0; n < N; n++) {
                    for (int x = S; x < image.length; x++) {
                        for (int y = T; y < image[0].length; y++) {
                            double real = 0;
                            double imaginary = 0;
```

```java
                    for (int s = 0; s < S; s++) {
                        for (int t = 0; t < T; t++) {
                            /*
                             * 影像與子小波的複共軛旋積
                             *
                             * childWavelets[s][t][m][n][0] 子小波的實部
                             * childWavelets[s][t][m][n][1] 子小波的虛部
                             * -childWavelets[s][t][m][n][1] 取共軛複數
                             */
                            real += image[x][y]
                                    * childWavelets[s][t][m][n][0];
                            imaginary += image[x][y]
                                    * -childWavelets[s][t][m][n][1];
                        }
                    }
                    this.gaborWavelet[x - S][y - T][m][n][0] = real;
                    this.gaborWavelet[x - S][y - T][m][n][1] =
                                                imaginary;

                }
            }
        }
    }

    // 用於計算母小波的函數 φ(x,y)
    // 傳回用複數表示（實部和虛部組成）的一維陣列 double[] {real,imaginary}
    private double[] computeMotherWavelet(double x, double y, int m,
int n) {

        double real = 1
                / (2 * Math.PI * sigmaX[m] * sigmaY[m])
                * Math.exp(-1
                / 2
```

```java
                      * (Math.pow(x, 2) / Math.pow(sigmaX[m], 2) + Math.pow(
                      y, 2) / Math.pow(sigmaY[m], 2)))
                      * Math.cos(2 * Math.PI * W[m] * x);
        double imaginary = 1
                      / (2 * Math.PI * sigmaX[m] * sigmaY[m])
                      * Math.exp(-1
                      / 2
                      * (Math.pow(x, 2) / Math.pow(sigmaX[m], 2) + Math.pow(
                      y, 2) / Math.pow(sigmaY[m], 2)))
                      * Math.sin(2 * Math.PI * W[m] * x);
        return new double[]{real, imaginary};
    }


    // x~
    private double xTilde(int x, int y, int m, int n) {
        return Math.pow(a, -m)
                      * (x * Math.cos(theta[n]) + y * Math.sin(theta[n]));
    }


    // y~
    private double yTilde(int x, int y, int m, int n) {
        return Math.pow(a, -m)
                      * (-x * Math.sin(theta[n] + y * Math.cos(theta[n])));
    }


    // 母小波經擴張和旋轉後產生的自相似 (self-similar) 的子小波 φmn
    // 傳回複數形式的子小波 double[] {real,imaginary}
    private double[] childWavelet(int x, int y, int m, int n) {
        double[] motherWavelet = computeMotherWavelet(xTilde(x, y, m, n),
                      yTilde(x, y, m, n), m, n);
        return new double[]{Math.pow(a, -m) * motherWavelet[0],
                      Math.pow(a, -m) * motherWavelet[1]};
    }
```

```java
// 根據 childWavelet 函數的計算公式計算出子小波
// 將結果的實部和虛部分別存入 double[][][][][0] 和 double[][][][][1]，傳回
private double[][][][][] computeChildWavelet() {
    double[][][][][] childWavelets = new double[S][T][M][N][2];
    double[] childWavelet;
    for (int s = 0; s < S; s++) {
        for (int t = 0; t < T; t++) {
            for (int m = 0; m < M; m++) {
                for (int n = 0; n < N; n++) {
                    childWavelet = childWavelet(s, t, m, n);
                    childWavelets[s][t][m][n][0] = childWavelet[0];
                    childWavelets[s][t][m][n][1] = childWavelet[1];
                }
            }
        }
    }
    return childWavelets;
}


// 計算強度
private double[][] computeMagnitudes(int[][] image) {
    double[][] magnitudes = new double[M][N];
    for (int i = 0; i < M; i++) {
        for (int j = 0; j < N; j++) {
            magnitudes[i][j] = 0;
        }
    }

    if (this.gaborWavelet == null) {
        computeGaborWavelet(image, computeChildWavelet());
    }
```

```java
    for (int m = 0; m < M; m++) {
        for (int n = 0; n < N; n++) {
            for (int x = S; x < image.length; x++) {
                for (int y = T; y < image[0].length; y++) {
                    magnitudes[m][n] += Math
                            .sqrt(Math
                                    .pow(this.gaborWavelet[x - S]
                                    [y - T][m][n][0], 2)
                                    + Math.pow(this.gaborWavelet
                                    [x - S][y - T][m][n][1], 2));
                }
            }
        }
    }
    return magnitudes;
}


// 對特徵值重新排序
public double[] normalize(double[] featureVector) {
    int dominantOrientation = 0;
    double orientationVectorSum = 0;
    double orientationVectorSum2 = 0;
    for (int m = 0; m < M; m++) {
        for (int n = 0; n < N; n++) {
            orientationVectorSum2 += Math.sqrt(Math.
                pow(featureVector[m * 2
                * N + n * 2], 2)
                + Math.pow(featureVector[m * 2 * N + n * 2 + 1], 2));
        }
        if (orientationVectorSum2 > orientationVectorSum) {
            orientationVectorSum = orientationVectorSum2;
            dominantOrientation = m;
        }
```

```java
        }

        double[] normalizedFeatureVector = new double[featureVector.
length];
        for (int m = dominantOrientation, k = 0; m < M; m++, k++) {
            for (int n = 0; n < N; n++) {
                normalizedFeatureVector[k * 2 * N + n * 2] =
                    featureVector[m * 2 * N + n * 2];
                normalizedFeatureVector[k * 2 * N + n * 2 + 1] =
                    featureVector[m * 2 * N + n * 2 + 1];
            }
        }
        for (int m = 0, k = M - dominantOrientation; m <
dominantOrientation; m++, k++) {
            for (int n = 0; n < N; n++) {
                normalizedFeatureVector[k * 2 * N + n * 2] =
                    featureVector[m * 2 * N + n * 2];
                normalizedFeatureVector[k * 2 * N + n * 2 + 1] =
                    featureVector[m * 2 * N + n * 2 + 1];
            }
        }

        return normalizedFeatureVector;
    }

    // Gabor特徵值
    public double[] computeGaborFeatures(String imageName) throws
IOException {
        File file = new File(imageName);
        BufferedImage image = ImageIO.read(file);
        image = ImageUtil.scaleImage(image, MAX_IMG_HEIGHT);
        WritableRaster raster = image.getRaster();
        int[][] grayLevel = new int[raster.getWidth()][raster.getHeight()];
```

```
int[] tmp = new int[3];
for (int i = 0; i < raster.getWidth(); i++) {
    for (int j = 0; j < raster.getHeight(); j++) {
        grayLevel[i][j] = raster.getPixel(i, j, tmp)[0];
    }
}
// 特徵值陣列 double[]{μ00,σ00,μ01,σ01,......, μMN, σMN}
double[] featureVector = new double[M * N * 2];
double[][] magnitudes = computeMagnitudes(grayLevel);
int imageSize = image.getWidth() * image.getHeight();
// σmn
double[][] sigmaMN = new double[M][N];


if (this.gaborWavelet == null) {
    computeGaborWavelet(grayLevel, computeChildWavelet());
}
// 計算特徵值 μmn 和 σmn
for (int m = 0; m < M; m++) {
    for (int n = 0; n < N; n++) {
        // μmn
        featureVector[m * 2 * N + n * 2] = magnitudes[m][n]
        / imageSize;
        for (int i = 0; i < sigmaMN.length; i++) {
            for (int j = 0; j < sigmaMN[0].length; j++) {
                sigmaMN[i][j] = 0.;
            }
        }
        for (int x = S; x < image.getWidth(); x++) {
            for (int y = T; y < image.getHeight(); y++) {
                sigmaMN[m][n] += Math.pow(
                        Math.sqrt(Math.pow(this.gaborWavelet
                            [x - S][y - T][m][n][0], 2)
                            + Math.pow(this.gaborWavelet
```

```
                                     [x - S][y - T][m][n][1], 2))
                              - featureVector[m * 2 * N + n * 2], 2);
                 }
           }

           featureVector[m * 2 * N + n * 2 + 1] = Math.
sqrt(sigmaMN[m][n])
                     / imageSize;
        }
    }
    this.gaborWavelet = null;

    return featureVector;
}

public String getGaborRepresentation(double[] featuresVector) {
    int vectorLength = featuresVector.length;
    StringBuilder sb = new StringBuilder(vectorLength);
    for (int i = 0; i < vectorLength; i++) {
        if(i==vectorLength-1) {
            sb.append(featuresVector[i]);
        } else {
            sb.append(featuresVector[i]+",");
        }
    }
    return sb.toString().trim();
}

public static void main(String args[]) {
    try {
        String imageName = "resource/image_name_rgb8.jpg";
        Gabor gabor = new Gabor();
        double[] featuresVector = gabor.computeGaborFeatures
```

```
(imageName);

System.out.println(gabor.getGaborRepresentation(gabor.normalize
(featuresVector)));
    } catch (IOException e) {
        // TODO Auto-generated catch block
        e.printStackTrace();
    }
  }
}
```

2.4.3 形狀特徵

影像中物體的形狀或是影像的區域形狀組成了影像的形狀特徵。物體形狀和區域形狀分別對應兩種形狀特徵表示方法 -- 輪廓特徵和區域特徵。輪廓特徵使用物體的外輪廓來表達，代表演算法是傅立葉描述符號。區域特徵使用影像整體的區域形狀表示，代表演算法是形狀不變矩。

由於輪廓特徵方法首先需要分析物體的輪廓，因此這裡介紹一下物體邊緣檢測演算法。Roberts、Sobel、Prewitt 和 Canny 是常用的邊緣檢測演算法，其中 Canny 演算法雖然較前 3 個演算法複雜，但檢測精度高，分析完整，抗噪能力好，也是我們通常使用的邊緣檢測演算法。

Canny 邊緣檢測演算法分 5 步驟，實作方式見程式 2-16。

步驟 1，降噪。使用高斯核心與影像進行旋積操作，去除影像雜訊。其中，二維高斯核心函數為：

$$G(x, y) = \frac{1}{2\delta\sigma^2} e^{-\frac{(x-x_c)^2+(y-y_c)^2}{2\sigma^2}} \tag{2-22}$$

步驟 2，尋找影像每個像素的梯度。首先使用 Roberts、Prewitt 和 Sobel 等某種邊緣檢測運算元，檢測橫縱兩個方向的梯度 G_x 和 G_y。由這些梯度可以計算出梯度模和梯度方向：

$$G = \sqrt{G_x{}^2 + G_y{}^2}$$
$$\theta = \arctan\left(\frac{G_y}{G_x}\right)$$

(2-23)

步驟 3，使用非最大值抑制排除假的邊緣點。非最大值抑制是一種邊緣細化技術，使在局部區域內不具有最大梯度值的像素點受到抑制。我們將梯度方向分為水平、垂直、主對角線方向、副對角線方向 4 個方向；計算每個像素點的梯度方向，並依據弧度值將其歸入這 4 個方向中的一種；然後將該像素的梯度模和這個方向上的鄰域像素的梯度模相比較，如果該像素的梯度模大，我們會將其納入邊緣點的考慮範圍，否則該像素點不計入邊緣點。

步驟 4，將在步驟 3 中納入邊緣考慮範圍的像素點使用落後設定值進行篩選。經過非最大值抑制後，選出的潛在邊緣點和真正的邊緣點已經很接近了。然而受雜訊和顏色變化的影響，仍然有某些「假」的邊緣點混入其中，可以採取使用高低兩個設定值的落後設定值來篩選邊緣點。如果梯度模大於高設定值，則將該像素標記為強邊緣點；如果梯度模小於高設定值而又大於低設定值，則將該像素標記為弱邊緣點；如果梯度模小於低設定值，該像素不計入邊緣點。

步驟 5，落後邊緣追蹤。步驟 4 中的強邊緣點一定會計入最後的邊緣點，但弱邊緣點可能是真正的邊緣點，也可能是雜訊或影像邊界。檢查已確定的邊緣點周圍的 8 個像素，與其相鄰的弱邊緣點是真正的邊緣點，不與其相鄰的弱邊緣點不是最後邊緣點。

⚈ 程式 2-16

```java
package com.ai.deepsearch.utils;

import java.awt.color.ColorSpace;
import java.awt.image.BufferedImage;
import java.awt.image.ColorConvertOp;
import java.awt.image.ConvolveOp;
import java.awt.image.Kernel;

/**
 * Canny 邊緣檢測
 */
public class CannyEdgeDetector {
    // 非邊緣點
    int[] noEdgePixel = {255};
    // 弱邊緣點
    int[] weakEdgePixel = {128};
    // 強邊緣點
    int[] strongEdgePixel = {0};
    int[] tmpPixel = {0};
    // 落後設定值
    // 低設定值
    private double thresholdLow = 60;
    // 高設定值
    private double thresholdHigh = 100;
    private BufferedImage image;

    public CannyEdgeDetector(BufferedImage image, double thresholdHigh,
                             double thresholdLow) {
        this.image = image;
        this.thresholdHigh = thresholdHigh;
        this.thresholdLow = thresholdLow;
```

```
    }

    public CannyEdgeDetector(BufferedImage image) {
        this.image = image;
    }


    // 產生高斯核心
    public float[] generateGaussianKernel(int radius, float sigma) {
        float center = (float) Math.floor((radius + 1) / 2);
        float[] kernel = new float[radius * radius];
        float sum = 0;
        for (int y = 0; y < radius; y++) {
            for (int x = 0; x < radius; x++) {
                int offset = y * radius + x;
                float distX = x - center;
                float distY = y - center;
                kernel[offset] = (float) ((1 / (2 * Math.PI * sigma *
sigma)) * Math
                        .exp(-(distX * distX + distY * distY)
                            / (2 * (sigma * sigma)))) ;
                sum += kernel[offset];
            }
        }
        // 歸一化
        for (int i = 0; i < kernel.length; i++)
            kernel[i] /= sum;
        return kernel;
    }

    // Canny 過濾，傳回的邊緣使用黑色表示，其他像素用白色表示
    public BufferedImage filter() {
        BufferedImage grayImage;
```

```java
// x 方向梯度
double[][] gx;
// y 方向梯度
double[][] gy;
// 梯度方向
double[][] gd;
// 梯度模
double[][] gm;

ColorConvertOp grayscale = new ColorConvertOp(
        ColorSpace.getInstance(ColorSpace.CS_GRAY), null);
grayImage = grayscale.filter(image, null);
// 1. 高斯模糊去噪
ConvolveOp gaussian = new ConvolveOp(new Kernel(5, 5,
        generateGaussianKernel(5, 1.4f)));
grayImage = gaussian.filter(grayImage, null);
// 2. 尋找影像梯度
// 利用 Sobel 梯度運算元求 x 和 y 方向的梯度
gx = sobelFilterX(grayImage);
gy = sobelFilterY(grayImage);
int width = grayImage.getWidth();
int height = grayImage.getHeight();
gd = new double[width][height];
gm = new double[width][height];
for (int x = 0; x < width; x++) {
    for (int y = 0; y < height; y++) {
        // 梯度方向 θ
        if (gx[x][y] != 0) {
            gd[x][y] = Math.atan(gy[x][y] / gx[x][y]);
        } else {
            gd[x][y] = Math.PI / 2d;
        }
```

```
        // 梯度模 G
        gm[x][y] = Math.sqrt(gy[x][y] * gy[x][y] + gx[x][y] *
gx[x][y]);
    }
}
// 3. 利用非最大值抑制排除假的邊緣點
// 4. 使用落後設定值進行篩選

// 影像四周邊界設為白色
for (int x = 0; x < width; x++) {
    grayImage.getRaster().setPixel(x, 0, noEdgePixel);
    grayImage.getRaster().setPixel(x, height - 1, noEdgePixel);
}
for (int y = 0; y < height; y++) {
    grayImage.getRaster().setPixel(0, y, noEdgePixel);
    grayImage.getRaster().setPixel(width - 1, y, noEdgePixel);
}

for (int x = 1; x < width - 1; x++) {
    for (int y = 1; y < height - 1; y++) {
        if (gd[x][y] < (Math.PI / 8d) && gd[x][y] >= (-Math.PI
/ 8d)) {
            // 像素 (x,y) 在水平區域的模最大
            if (gm[x][y] > gm[x + 1][y] && gm[x][y] > gm[x - 1][y])
                // 潛在邊緣點
                setPixel(x, y, grayImage, gm[x][y]);
            else
                // 非邊緣點
                grayImage.getRaster().setPixel(x, y, noEdgePixel);
        } else if (gd[x][y] < (3d * Math.PI / 8d)
                && gd[x][y] >= (Math.PI / 8d)) {
            // 像素 (x,y) 在主對角線區域的模最大
```

```
                    if (gm[x][y] > gm[x - 1][y - 1]
                            && gm[x][y] > gm[x + 1][y + 1])
                        // 潛在邊緣點
                        setPixel(x, y, grayImage, gm[x][y]);
                    else
                        // 非邊緣點
                        grayImage.getRaster().setPixel(x, y, noEdgePixel);
            } else if (gd[x][y] < (-3d * Math.PI / 8d)
                    || gd[x][y] >= (3d * Math.PI / 8d)) {
                // 像素 (x,y) 在垂直區域的模最大
                if (gm[x][y] > gm[x][y + 1] && gm[x][y] > gm[x][y - 1])
                    // 潛在邊緣點
                    setPixel(x, y, grayImage, gm[x][y]);
                else
                    // 非邊緣點
                    grayImage.getRaster().setPixel(x, y, noEdgePixel);
            } else if (gd[x][y] < (-Math.PI / 8d)
                    && gd[x][y] >= (-3d * Math.PI / 8d)) {
                // 像素 (x,y) 在副對角線區域的模最大
                if (gm[x][y] > gm[x + 1][y - 1]
                        && gm[x][y] > gm[x - 1][y + 1])
                    // 潛在邊緣點
                    setPixel(x, y, grayImage, gm[x][y]);
                else
                    // 非邊緣點
                    grayImage.getRaster().setPixel(x, y, noEdgePixel);
            } else {
                // 非邊緣點
                grayImage.getRaster().setPixel(x, y, noEdgePixel);
            }
        }
    }
```

```java
        // 5. 落後邊緣追蹤
        int[] tmpArray = {0};
        for (int x = 1; x < width - 1; x++) {
            for (int y = 1; y < height - 1; y++) {
                if (grayImage.getRaster().getPixel(x, y, tmpArray)[0] <
50) {

                    // 追蹤強邊緣點 (x,y) 周圍的弱邊緣點
                    trackWeakOnes(x, y, grayImage);
                }
            }
        }
        // 去除單一的弱邊緣點
        for (int x = 2; x < width - 2; x++) {
            for (int y = 2; y < height - 2; y++) {
                if (grayImage.getRaster().getPixel(x, y, tmpArray)[0] >
50) {

                    grayImage.getRaster().setPixel(x, y, noEdgePixel);
                }
            }
        }
        return grayImage;
    }


    // 遞迴追蹤強邊緣點周圍的弱邊緣點
    private void trackWeakOnes(int x, int y, BufferedImage grayImage) {
        for (int xx = x - 1; xx <= x + 1; xx++)
            for (int yy = y - 1; yy <= y + 1; yy++) {
                // 點 (x,y) 周圍的 8 個相鄰點是否是弱邊緣點
                if (isWeak(xx, yy, grayImage)) {
                    grayImage.getRaster().setPixel(xx, yy,
strongEdgePixel);
                    trackWeakOnes(xx, yy, grayImage);
```

```
                }
            }
        }

    // 判斷是否是弱邊緣點
    private boolean isWeak(int x, int y, BufferedImage grayImage) {
        return (grayImage.getRaster().getPixel(x, y, tmpPixel)[0] > 0
&& grayImage
                .getRaster().getPixel(x, y, tmpPixel)[0] < 255);
    }

    // 依據落後設定值的高設定值和低設定值來區分潛在邊緣點中的弱邊緣點和強邊緣點
    private void setPixel(int x, int y, BufferedImage grayImage,
double v) {
        if (v > thresholdHigh)
            grayImage.getRaster().setPixel(x, y, strongEdgePixel);
        else if (v > thresholdLow)
            grayImage.getRaster().setPixel(x, y, weakEdgePixel);
        else
            grayImage.getRaster().setPixel(x, y, noEdgePixel);
    }

    // 與水平 Sobel 濾波運算元旋積
    private double[][] sobelFilterX(BufferedImage grayImage) {
        double[][] result = new double[grayImage.getWidth()][grayImage
                .getHeight()];
        int[] tmpArray = new int[1];
        int sum;
        for (int x = 1; x < grayImage.getWidth() - 1; x++) {
            for (int y = 1; y < grayImage.getHeight() - 1; y++) {
                sum = 0;
                sum += grayImage.getRaster().getPixel(x - 1, y - 1,
```

```
tmpArray)[0];
                sum += 2 * grayImage.getRaster().getPixel(x - 1, y,
tmpArray)[0];
                sum += grayImage.getRaster().getPixel(x - 1, y + 1,
tmpArray)[0];
                sum -= grayImage.getRaster().getPixel(x + 1, y - 1,
tmpArray)[0];
                sum -= 2 * grayImage.getRaster().getPixel(x + 1, y,
tmpArray)[0];
                sum -= grayImage.getRaster().getPixel(x + 1, y + 1,
tmpArray)[0];
                result[x][y] = sum;
            }
        }
        for (int x = 0; x < grayImage.getWidth(); x++) {
            result[x][0] = 0;
            result[x][grayImage.getHeight() - 1] = 0;
        }
        for (int y = 0; y < grayImage.getHeight(); y++) {
            result[0][y] = 0;
            result[grayImage.getWidth() - 1][y] = 0;
        }
        return result;
    }


    // 與垂直 Sobel 濾波運算元旋積
    private double[][] sobelFilterY(BufferedImage gray) {
        double[][] result = new double[gray.getWidth()][gray.getHeight()];
        int[] tmpArray = new int[1];
        int sum = 0;
        for (int x = 1; x < gray.getWidth() - 1; x++) {
            for (int y = 1; y < gray.getHeight() - 1; y++) {
```

```
            sum = 0;
            sum += gray.getRaster().getPixel(x - 1, y - 1,
tmpArray)[0];
            sum += 2 * gray.getRaster().getPixel(x, y - 1,
tmpArray)[0];
            sum += gray.getRaster().getPixel(x + 1, y - 1,
tmpArray)[0];
            sum -= gray.getRaster().getPixel(x - 1, y + 1,
tmpArray)[0];
            sum -= 2 * gray.getRaster().getPixel(x, y + 1,
tmpArray)[0];
            sum -= gray.getRaster().getPixel(x + 1, y + 1,
tmpArray)[0];
            result[x][y] = sum;
        }
    }
    for (int x = 0; x < gray.getWidth(); x++) {
        result[x][0] = 0;
        result[x][gray.getHeight() - 1] = 0;
    }
    for (int y = 0; y < gray.getHeight(); y++) {
        result[0][y] = 0;
        result[gray.getWidth() - 1][y] = 0;
    }
    return result;
    }
}
```

1 傅立葉形狀描述符號

傅立葉形狀描述符號是對影像中物體形狀的輪廓線進行傅立葉轉換，由獲得的傅立葉係數組成向量來對影像進行代表。傅立葉形狀描述符號表示的是影像形狀的頻域特徵，具有較好的抗噪特性，並進一步降低了描述符號對邊界變化的敏感度。Granlund 在 1972 年提出了一種具有不變性的傅立葉形狀描述符號，用於手寫字元識別，並取得 98% 的識別正確率。

$$FD_k = \frac{\sqrt{a_{xk}^2 + a_{yk}^2}}{\sqrt{a_{x1}^2 + a_{y1}^2}} + \frac{\sqrt{b_{xk}^2 + b_{yk}^2}}{\sqrt{b_{x1}^2 + b_{y1}^2}}$$
(2-24)

我們可以將影像輪廓（如圖 2-11 所示）視為在一個複平面上，那麼輪廓中的每個點就可以用一個複數來表示。整個影像輪廓就轉化為一個複數序列 $c(t)=x(t)+jy(t)$，其中 $t=0, 1, \cdots, N-1$。對該序列進行傅立葉轉換後可以獲得它的傅立葉係數 a_{xk} 和 b_{xk}，及其複共軛的傅立葉係數 a_{yk} 和 b_{yk}。進一步利用式（2-24），我們就可以求得傅立葉描述符號 FD_k。

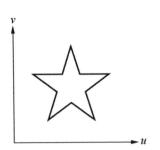

圖 2-11 複平面上的影像輪廓

研究人員發現 10 個傅立葉係數已經能夠極佳地表示形狀，故傅立葉係數數量常預設為 10，實作方式如程式 2-17[7]。

7　Zhang D, Lu G. Review of shape representation and description techniques[J]. Pattern Recognition, 2004, 37(1):1-19.

☒ 程式 2-17

```java
package com.ai.deepsearch.features.global.shape;

import com.ai.deepsearch.utils.CannyEdgeDetector;

import javax.imageio.ImageIO;
import java.awt.image.BufferedImage;
import java.io.File;
import java.io.IOException;
import java.util.ArrayList;
import java.util.List;

// 像素點類別
class Point implements Comparable<Point> {
    public int x;
    public int y;

    public Point(int x, int y) {
        this.x = x;
        this.y = y;
    }

    public String toString() {
        return "x: " + this.x + " y: " + this.y;
    }

    @Override
    public int compareTo(Point point) {
        int result = Integer.compare(this.x, point.x);
        if (result != 0) {
            return result;
        }
```

```
        return Integer.compare(this.y, point.y);
    }
}


/**
 * 傅立葉形狀描述符號
 */
public class FourierShapeDescriptor {
    private static final int DEFAULT_FOURIER_COEFF_NUM = 10;
    private double[] ax;
    private double[] bx;
    private double[] ay;
    private double[] by;
    private double[] efd;

    // 計算傅立葉轉換後的係數
    private double axCoefficient(Point[] points, int k) {
        double ax = 0d;
        int contourNum = points.length;
        for (int i = 0; i < contourNum; i++) {
            ax += points[i].x * Math.cos(2 * k * Math.PI * i / contourNum);
        }
        return ax * 2 / contourNum;
    }


    private double bxCoefficient(Point[] points, int k) {
        double bx = 0d;
        int contourNum = points.length;
        for (int i = 0; i < contourNum; i++) {
            bx += points[i].x * Math.sin(2 * k * Math.PI * i / contourNum);
        }
        return bx * 2 / contourNum;
```

```java
        }

    private double ayCoefficient(Point[] points, int k) {
        double ay = 0d;
        int contourNum = points.length;
        for (int i = 0; i < contourNum; i++) {
            ay += points[i].y * Math.cos(2 * k * Math.PI * i / contourNum);
        }
        return ay * 2 / contourNum;
    }

    private double byCoefficient(Point[] points, int k) {
        double by = 0d;
        int contourNum = points.length;
        for (int i = 0; i < contourNum; i++) {
            by += points[i].y * Math.sin(2 * k * Math.PI * i / contourNum);
        }
        return by * 2 / contourNum;
    }

    // 使用 Canny 演算法取得影像輪廓點
    private List<Point> getEdgePoints(BufferedImage image) {
        CannyEdgeDetector cannyEdgeDetector = new CannyEdgeDetector(image);
        BufferedImage edgeImage = cannyEdgeDetector.filter();
        int width = edgeImage.getWidth();
        int height = edgeImage.getHeight();
        List<Point> points = new ArrayList<>();
        int[] tmpArray = new int[1];
        for (int x = 0; x < width; x++) {
            for (int y = 0; y < height; y++) {
                if (edgeImage.getRaster().getPixel(x, y, tmpArray)[0]
== 0) {
```

```java
                    points.add(new Point(x,y));
                }
            }
        }
        return points;
    }


    private void getFourierShapeDescriptor(Point[] points, int fdNum) {
        this.ax = new double[fdNum];
        this.bx = new double[fdNum];
        this.ay = new double[fdNum];
        this.by = new double[fdNum];

        for (int k = 0; k < fdNum; k++) {
            this.ax[k] = axCoefficient(points, k);
            this.bx[k] = bxCoefficient(points, k);
            this.ay[k] = ayCoefficient(points, k);
            this.by[k] = byCoefficient(points, k);
        }


        this.efd = new double[fdNum];

        for (int k = 0; k < fdNum; k++) {
            efd[k] = Math.sqrt((this.ax[k] * this.ax[k] + this.ay[k]
                    * this.ay[k])
                    / (this.ax[1] * this.ax[1] + this.ay[1] * this.ay[1]))
                    + Math.sqrt((this.bx[k] * this.bx[k] + this.by[k]
                    * this.by[k])
                    / (this.bx[1] * this.bx[1] + this.by[1]
                    * this.by[1]));
        }
    }
```

```java
    public String getFourierSDRepresentation() {
        StringBuilder sb = new StringBuilder(DEFAULT_FOURIER_COEFF_NUM);
        for (int k = 0; k < DEFAULT_FOURIER_COEFF_NUM; k++) {
            if(k==DEFAULT_FOURIER_COEFF_NUM-1){
                sb.append(String.valueOf(this.efd[k]));
            } else {
                sb.append(String.valueOf(this.efd[k])+",");
            }
        }
        return sb.toString().trim();
    }


    public void computeFourierShapeDescriptor(String imageName)
            throws IOException {
        File file = new File(imageName);
        BufferedImage image = ImageIO.read(file);
        List<Point> pointsList = getEdgePoints(image);
        Point[] pointsArray=pointsList.toArray(new Point[pointsList.
size()]);
        getFourierShapeDescriptor(pointsArray, DEFAULT_FOURIER_COEFF_NUM);
    }


    public static void main(String[] args) {
        try {
            String imageName = "resource/image_name_rgb8.jpg";
            FourierShapeDescriptor fourierSD = new
FourierShapeDescriptor();
            fourierSD.computeFourierShapeDescriptor(imageName);
            System.out.println(fourierSD.getFourierSDRepresentation());
        } catch (IOException e) {
            // TODO Auto-generated catch block
```

```
            e.printStackTrace();
        }
    }
}
```

2 形狀不變矩

矩是物理學中的概念，由 Pearson 在 1894 年引入統計學，它通常用來表示隨機變數的分佈情況。設 X 為隨機變數，若 $E(X)$ 存在，且 $E(|X-E(X)|^k)<+\infty$，則稱 $\mu_k(X)=E[(X-E(X))^k]$ 為 X 的 k 階中心矩。1 階中心矩為原點，2 階中心矩為 X 的方差，3 階中心矩為 X 的偏度，4 階中心矩為 X 的峰度。對一幅影像來說，矩表現了影像灰階的分佈狀況，它可以代表一幅影像的幾何特徵。

1962 年，MING-KUEI HU 在他的論文中複述了中心矩、$p+q$ 階矩等概念，並在此基礎上建置了具有轉換、旋轉、縮放獨立性的 7 個矩。

一個二維影像的 $p+q$ 階原點矩可表示為（其中 M、N 表示影像的寬和高，$f(x, y)$ 為灰階分佈函數）：

$$m_{pq} = \sum_{x=1}^{M}\sum_{y=1}^{N} x^p y^q f(x,y) \tag{2-25}$$

中心矩可表示為：

$$\mu_{pq} = \sum_{x=1}^{M}\sum_{y=1}^{N} (x-\overline{x})^p (y-\overline{y})^q f(x,y) \tag{2-26}$$

可對中心矩進行歸一化，以使其獲得縮放獨立性：

$$\eta_{pq} = \frac{\mu_{pq}}{\mu_{00}^{\gamma}}, \gamma = \frac{p+q+2}{2} \tag{2-27}$$

HU 以以上結論為基礎提出的 7 個不變矩，如式（2-28）所示：

$$\phi_1 = \mu_{20} + \mu_{02}$$

$$\phi_2 = (\mu_{20} - \mu_{02})^2 + 4\mu_{11}^{\ 2}$$

$$\phi_3 = (\mu_{30} - 3\mu_{12})^2 + (3\mu_{21} - \mu_{03})^2$$

$$\phi_4 = (\mu_{30} + \mu_{12})^2 + (\mu_{21} + \mu_{03})^2$$

$$\phi_5 = (\mu_{30} - 3\mu_{12})(\mu_{30} + \mu_{12})\left[(\mu_{30} + \mu_{12})^2 - 3(\mu_{21} + \mu_{03})^2\right]$$
$$\qquad + (3\mu_{21} - 3\mu_{03})(\mu_{21} + \mu_{03})\left[(3\mu_{30} + \mu_{12})^2 - (\mu_{21} + \mu_{03})^2\right]$$

$$\phi_6 = (\mu_{20} - \mu_{02})\left[(\mu_{30} + \mu_{12})^2 - (\mu_{21} + \mu_{03})^2\right] + 4\mu_{11}(\mu_{30} + \mu_{12})(\mu_{21} + \mu_{03})$$

$$\phi_7 = 3(\mu_{21} - \mu_{03})(\mu_{30} + \mu_{12})\left[(\mu_{30} + \mu_{12})^2 - 3(\mu_{21} + \mu_{03})^2\right]$$
$$\qquad - (\mu_{30} - 3\mu_{21})(\mu_{21} + \mu_{03})\left[3(\mu_{30} - \mu_{12})^2 - (\mu_{21} + \mu_{03})^2\right]$$

(2-28)

實作方式見程式 2-18。

⧗ 程式 2-18

```java
package com.ai.deepsearch.features.global.shape;

import com.ai.deepsearch.utils.ImageUtil;

import javax.imageio.ImageIO;
import java.awt. *;
import java.awt.color.ColorSpace;
import java.awt.image.BufferedImage;
import java.awt.image.ColorConvertOp;
import java.awt.image.WritableRaster;
import java.io.File;
import java.io.IOException;

/**
 * 形狀不變矩
 */
```

```java
public class ShapeInvariantMoments {
    private static final int MAX_IMG_HEIGHT = 64;
    private int[][] grayMatrix;
    private double xBar = 0;
    private double yBar = 0;
    private double[] moments;

    // p+q 階矩
    private double pqMoment(int p, int q) {
        float m = 0;
        for (int x = 0; x < this.grayMatrix.length; x++) {
            for (int y = 0; y < this.grayMatrix[0].length; y++) {
                m += Math.pow(x, p) * Math.pow(y, q) * this.
grayMatrix[x][y];
            }
        }
        return m;
    }

    // 中心矩
    private double centralMoment(int p, int q) {
        float cm = 0;
        for (int x = 0; x < this.grayMatrix.length; x++) {
            for (int y = 0; y < this.grayMatrix[0].length; y++) {
                cm += Math.pow(x - this.xBar, p) * Math.pow(y -
                    this.yBar, q) * this.grayMatrix[x][y];
            }
        }
        return cm;
    }

    // 歸一化的中心矩
    private double mu(int p, int q) {
```

```java
        float gamma = (p + q) / 2 + 1;
        return centralMoment(p, q) / Math.pow(centralMoment(0, 0),
gamma);
    }


    public String getMomentsRepresentation(double[] moments) {
        StringBuilder sb = new StringBuilder(moments.length);
        for (int i = 0; i < moments.length; i++) {
            if(i==moments.length-1) {
                sb.append(moments[i]);
            } else {
                sb.append(moments[i]+",");
            }
        }
        return sb.toString().trim();
    }


    // 計算形狀不變矩
    public void computeShapeInvariantMoments(String imageName)
            throws IOException {
        File file = new File(imageName);
        BufferedImage image = ImageIO.read(file);
        // 轉為灰階色彩空間
        ColorConvertOp colorConvertOp = new ColorConvertOp(image
                .getColorModel().getColorSpace(),
                ColorSpace.getInstance(ColorSpace.CS_GRAY),
new RenderingHints(
                RenderingHints.KEY_COLOR_RENDERING,
                RenderingHints.VALUE_COLOR_RENDER_QUALITY));
        BufferedImage grayImage = colorConvertOp.filter(image, null);
        // 統一影像大小
        grayImage = ImageUtil.scaleImage(grayImage, MAX_IMG_HEIGHT);
        WritableRaster raster = grayImage.getRaster();
```

```java
int width = grayImage.getWidth();
int height = grayImage.getHeight();
int[] pixel = new int[3];
this.grayMatrix = new int[width][height];
for (int x = 0; x < width; x++) {
    for (int y = 0; y < height; y++) {
        raster.getPixel(x, y, pixel);
        grayMatrix[x][y] = pixel[0];
    }
}
double m00 = pqMoment(0, 0);


this.xBar = pqMoment(1, 0) / m00;
this.yBar = pqMoment(0, 1) / m00;
this.moments = new double[7];
// μ20+μ02
this.moments[0] = mu(2, 0) + mu(0, 2);
// (μ20-μ02)2+4μ112
this.moments[1] = Math.pow(mu(2, 0) - mu(0, 2), 2) + 4
        * Math.pow(mu(1, 1), 2);
// (μ30-3μ12)2+(3μ21-μ03)2
this.moments[2] = Math.pow(mu(3, 0) - 3 * mu(1, 2), 2)
        + Math.pow(3 * mu(2, 1) - mu(0, 3), 2);
// (μ30+μ12)2+(μ21+μ03)2
this.moments[3] = Math.pow(mu(3, 0) + mu(1, 2), 2)
        + Math.pow(mu(2, 1) + mu(0, 3), 2);
// (μ30-3μ12)(μ30+μ12)[(μ30+μ12)2-3(μ21+μ03)2]+(3μ21-μ03)
(μ21+μ03)[3(μ30+μ12)2
// -(μ21+μ03)2]
this.moments[4] = (mu(3, 0) - 3 * mu(1, 2))
        * (mu(3, 0) + mu(1, 2))
        * (Math.pow(mu(3, 0) + mu(1, 2), 2) - 3 * Math.
pow(mu(2, 1)
```

```java
                    + mu(0, 3), 2))
                    + (3 * mu(2, 1) - mu(0, 3))
                    * (mu(2, 1) + mu(0, 3))
                    * (3 * Math.pow(mu(3, 0) + mu(1, 2), 2) - Math.
pow(mu(2, 1)
                    + mu(0, 3), 2));
            // (μ20-μ02)[(μ30+μ12)2-(μ21+μ03)2]+4μ11(μ30+μ12)(μ21+μ03)
            this.moments[5] = (mu(2, 0) - mu(0, 2))
                    * (Math.pow(mu(3, 0) + mu(1, 2), 2) - Math.pow(
                    mu(2, 1) + mu(0, 3), 2)) + 4 * mu(1, 1)
                    * (mu(3, 0) + mu(1, 2)) * (mu(2, 1) + mu(0, 3));
            // (3μ21-μ03)(μ30+μ12)[(μ30+μ12)2-3(μ21+μ03)2]-(μ30-3μ12)
(μ21+μ03)[3(μ30+μ12)2
            // -(μ21+μ03)2]
            this.moments[6] = (3 * mu(2, 1) - mu(0, 3))
                    * (mu(3, 0) + mu(1, 2))
                    * (Math.pow(mu(3, 0) + mu(1, 2), 2) - 3 * Math.
pow(mu(2, 1)
                    + mu(0, 3), 2))
                    - (mu(3, 0) - 3 * mu(1, 2))
                    * (mu(2, 1) + mu(0, 3))
                    * (3 * Math.pow(mu(3, 0) + mu(1, 2), 2) - Math.
pow(mu(2, 1)
                    + mu(0, 3), 2));
    }

    public static void main(String[] args) {
        try {
            String imageName = "resource/image_name_rgb8.jpg";
            ShapeInvariantMoments huMoments = new
ShapeInvariantMoments();
            huMoments.computeShapeInvariantMoments(imageName);
```

```
            System.out.println(huMoments
                    .getMomentsRepresentation(huMoments.moments));
        } catch (IOException e) {
            // TODO Auto-generated catch block
            e.printStackTrace();
        }
    }
}
```

2.5 局部特徵

如 2.3 節所述，每幅影像都含有局部上的特徵。這些特徵可以是一朵小
花、一棵小草、一棟小房子。特徵區域常常與其周圍具有明顯的顏色或
灰階上的差別。影像研究學者使用數學的方法對這些區域進行特徵分析
和表達，形成了 SIFT、SURF 等具有縮放、旋轉，甚至仿射，以及光源
改變不變性的局部特徵描述符號。下面我們將逐一介紹 SIFT 和 SURF
的原理以及實現方法。由於 SIFT 和 SURF 的作者都對其演算法申請了
專利，所以在此不再提供相關演算法實現的原始程式碼。

2.5.1 SIFT 描述符號

1999 年，加拿大英屬哥倫比亞大學的 David G.Lowe 教授在電腦視覺國
際會議（ICCV）上第一次提出了以尺度空間為基礎的影像局部特徵描述
符號[8]（Scale Invariant Feature Transform，SIFT）。它不僅具有對影像平

8　Lowe D G. Object Recognition from Local Scale-Invariant Features[C]//iccv.IEEE Computer
　　Society,1999:1150.

移、縮放、旋轉的不變性,而且具有仿射、投影轉換的不變性,甚至在不同光源條件下具有不變性。

檢測局部特徵點通常採用高斯拉普拉斯(Laplace of Gaussian,LoG)或赫森行列式(Determinant of Hessian,DoH)方法。Laplace 運算元可以用來檢測影像中的局部極值點,但它無法極佳地應對影像中的雜訊,而這剛好是高斯函數的強項。我們首先使用高斯核心與影像進行旋積,以達到清除影像雜訊的目的。

$$L(x,y,\sigma)=I(x,y)*G(x,y,\sigma) \tag{2-29}$$

然後對去除雜訊後的影像進行拉普拉斯轉換。

$$\nabla^2\left[I(x,y)*G(x,y,\sigma)\right] = \nabla^2\left[G(x,y,\sigma)\right]*I(x,y) \tag{2-30}$$

根據公式,我們可以先對高斯核心進行拉普拉斯轉換,再與影像進行旋積運算,也就是利用高斯拉普拉斯運算元(LoG)來檢測局部特徵點。當 LoG 的尺度與影像中某個特徵點的尺度相同時,LoG 才會產生較強的回應,所以對於影像中的諸多特徵點,需要在不同的尺度條件下才能檢測出來。建置連續的尺度空間用於檢測特徵點成為了一種可行的解決方案。由於 LoG 的計算量很大,Lowe 在 SIFT 演算法中使用計算量小且與 LoG 近似的高斯函數差分(Difference of Gaussian,DoG)來檢測尺度空間的極值點。

SIFT 演算法分為以下 4 個步驟。

1 檢測尺度空間極值點

改變高斯函數 $G(x,y,\sigma)=\dfrac{1}{2\pi\sigma^2}e^{-(x^2+y^2)/2\sigma^2}$ 中的尺度變數 σ,並與影像 $I(x,y)$ 進行旋積運算,將旋積後的一系列影像用於建置連續的尺度空間。沿著尺度增大的方向，以初始尺度 σ_0 的 2^n 倍作為起始值，將尺度空間分為許多部分，每個部分被稱為一個 Octave。每個 Octave 組中影像大小相同，下一組的影像是上一組影像的降取樣，長寬各是原影像的一半。Octave 各組影像逐層降取樣，組成了類似金字塔形狀的結構，如圖 2-12 所示。金字塔的層數由原始影像的大小和塔頂影像的大小共同決定，$n_{octave} = \log_2\left[\min(M_{init}, N_{init})\right] - \log_2\left[\min(M_{top}, N_{top})\right]$。由於採用高斯函數平滑原始影像會在其高頻部分產生部分損失，Lowe 在論文中建議首先將原始影像的長寬各擴充一倍，並進行線性內插，將其作為起始影像。由於影像在照相時，鏡頭已經對它進行一定量的模糊，Lowe 將其值設為 0.5，初始尺度 σ_0 設定為 1.6，實際作用在原始影像上的尺度為 $\sqrt{1.6^2 - (2\times 0.5)^2}$。沿著尺度增大的方向，每個 Octave 又取樣為 s 層（intervals），s 一般為 3 ～ 5，Octave 中的影像大小相同，尺度按照 $k^0\sigma_0$，$k^1\sigma_0$，$k^2\sigma_0$，$k^3\sigma_0$，…，$k^s\sigma_0$ 的規律增加，其中 k^s=2。由於在尺度空間極值比較過程中，每個 Octave 中的最上層和最下層影像都沒有相鄰影像，因此無法進行極值比較。我們在高斯空間（Gaussian Space）中繼續按照尺度的變化規律進行高斯模糊，在其頂層上又產生了 3 幅影像，就有了 $s+3$ 層影像。由於 DoG 空間（DoG Space）中的影像是高斯空間中相鄰影像相減而成，那麼在其中就有 $s+2$ 層影像。

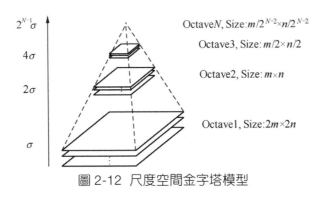

圖 2-12 尺度空間金字塔模型

圖 2-13 示範了當 s=3 時尺度空間的建置情況。在高斯空間中有 6 層影像，在 DoG 空間中有 5 層影像。圖中虛線箭頭說明上一組 Octave 中倒數第二層影像隔點取樣產生了下一組 Octave 中的底層影像，進而保持了尺度的連續性。

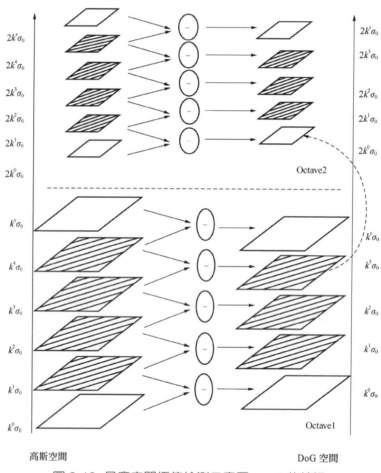

圖 2-13 尺度空間極值檢測示意圖，s=3 的情況

為了在 DoG 空間中檢測極值點，需要將影像中每個像素點的灰階值與其周圍 8 個點以及 DoG 空間中上下兩層相鄰影像中的各 9 個相鄰點，共計

26（8+9×2）個點的灰階值進行比較，如圖 2-14 所示。如果這個像素點的灰階值在這 27 個點中是最大或最小值，那麼此點就是該尺度下的極值點。

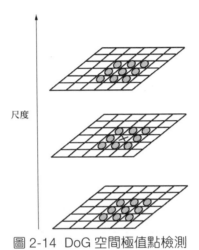

尺度

圖 2-14 DoG 空間極值點檢測

2 在極值點中定位出更加穩定的關鍵點

由於低比較度的極值點對於雜訊更加敏感，因此我們首先需要去除低比較度的極值點。DoG 函數在影像邊緣會產生較強的邊緣回應，所以還需要去除不穩定的邊緣回應點。極值點經過以上兩步驟的處理後便篩選出了更加穩定的關鍵點。

3 關鍵點方向分配

在連續尺度空間上取得的一系列關鍵點具有縮放不變的性質。下面來看一看怎樣才能使它具有旋轉不變性呢？我們需要結合影像的局部特徵，給每個關鍵點分配一個方向。對於在 DoG 空間中檢測出的關鍵點，按照式（2-31）計算對應尺度空間 L 高斯模糊後的影像 3σ 鄰域視窗內像素的梯度和方向分佈情況。

$$m(x, y) = \sqrt{(L(x+1, y) - L(x-1, y))^2 + (L(x, y+1) - L(x, y-1))^2}$$

$$\theta(x, y) = \arctan\left(\frac{L(x, y+1) - L(x, y-1)}{L(x+1, y) - L(x-1, y)}\right) \tag{2-31}$$

將計算出的鄰域內像素點的梯度及方向使用長條圖進行統計。長條圖將 360° 方向分為 36 個區間（bins），每個區間的範圍為 10°。如圖 2-15 所示，圖中為方便說明將 36 個柱簡化為 8 個，將該長條圖中梯度最大值的方向作為關鍵點的主方向。為增強符合的穩定性，將大於最大值 80% 的梯度方向作為輔方向。

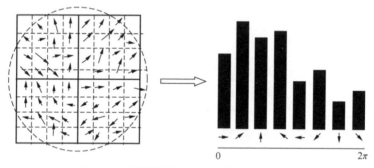

圖 2-15 關鍵點方向長條圖產生示意圖

■4 關鍵點特徵描述

現在我們需要使用在前 3 步驟中獲得的尺度、方向和位置，為每個關鍵點建立一個具有各種不變性的特徵描述符號。該特徵描述符號實際上是在步驟 3 所產生的梯度長條圖的一種數學向量表達。

（1）確定關鍵點周圍的影像區域。在方向長條圖中計算了以關鍵點為圓心，半徑 $r = \dfrac{3\sigma_{oct} \times \sqrt{2} \times (d+1)}{2}$（$\sigma_{oct}$ 為 octave 組內影像尺度，$d=4$）的圓形區域內像素梯度方向分佈情況。

（2）為確保描述符號的旋轉不變性，將座標軸旋轉為和主方向保持一致。

（3）計算圓形區域 16×16 視窗劃分後每個像素的梯度，並使用 $\sigma =0.5d$
的高斯函數加權（d =4）。然後每 4×4 個小格組成一個子區域，共
4×4 個子區域，將子區域內像素的梯度加權累加到 8 個方向上，如
圖 2-16 所示。這樣每個關鍵點就產生了一個 4×4×8=128 維的描
述向量。為使該描述符號具有光源不變性，將每一維數值除以 128
維數值和的平方根進行歸一化。這個 128 維的描述向量就表現了影
像的某個局部特徵。

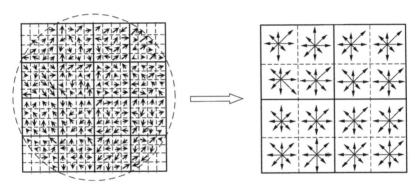

圖 2-16　4×4 區域 8 方向梯度

2.5.2　SURF 描述符號

SIFT 描述符號雖然使用 DoG 作為 LoG 的近似而進一步簡化計算，但其
計算量依然很大。SIFT 檢測特徵點多、效能穩定，但其複雜度較高，
需要消耗大量的時間和運算資源。以 SIFT 為基礎的這些問題，2006 年
Herbert Bay 提出加速穩健特徵（Speed-Up Robust Features，SURF）[9] 演
算法。該演算法在 SIFT 研究的基礎上進行改進，採用了與 SIFT 相近的

9　Bay H, Tuytelaars T, Gool L V. SURF: speeded up robust features[C]// European Conference on Computer Vision. Springer-Verlag, 2006:404-417.

思維和步驟：首先檢測尺度空間極值點，然後定位關鍵點，接著對關鍵點進行方向分配，最後產生關鍵點的向量描述符號。SURF 對 SIFT 中的多個方法進行了改進和簡化，相當大地縮短了產生影像局部特徵描述的時間。下面來分析一下 SURF 在哪些方面做出了改進和簡化。

在 2.5.1 節提到了檢測局部特徵點常用的兩種方法 LoG 和 DoH，SIFT 使用了 LoG 的近似 DoG，而 SURF 使用 DoH 的近似。當尺度為 σ 時，影像中的點 $P(x, y)$ 的 Hessian 矩陣 $H(P,\sigma)=\begin{bmatrix} L_{xx}(P,\sigma) & L_{xy}(P,\sigma) \\ L_{xy}(P,\sigma) & L_{yy}(P,\sigma) \end{bmatrix}$，$L_{xx}(P,\sigma)$ 代表二維高斯函數 $g(x, y, \sigma)$ 關於 x 的二階偏導 $\dfrac{\partial^2 g(x,y,\sigma)}{\partial x^2}$ 與影像在點 P 處旋積的結果，$L_{xy}(P, \sigma)$ 代表二維高斯函數 $g(x,y,\sigma)$ 先後求 x 和 y 的偏導 $\dfrac{\partial^2 g(x,y,\sigma)}{\partial x \partial y}$ 與影像在點 P 處旋積的結果 $L_{yy}(P, \sigma)$ 與 $L_{xx}(P, \sigma)$ 計算方式相同。DoH 是 Hessian 矩陣的行列式：$det(H)=L_{xx}(P,\sigma)L_{yy}(P,\sigma)-L_{xy}^2(P,\sigma)$，它可以判斷點 P 是否為極值點。Herbert Bay 在 SURF 論文中使用盒子濾波器作為離散二維高斯函數的二階偏導的近似，相當大地加強了運算速度。如圖 2-17 所示，左邊由上到下分別是 x 方向、y 方向、xy 方向的離散二維高斯函數的二階偏導數，右邊則分別是它們的近似，9×9 的盒子濾波器近似相等於 $\sigma =1.2$ 的二維高斯函數的二階偏導數。

Herbert Bay 還使用一種叫作「積分圖」的運算方式，它將盒子濾波器與每個像素的旋積運算轉化為對「積分圖」的加減運算，進一步使其消耗更小，速度更快。積分圖 $I_\Sigma(P)$ 在點 $P(x, y)$ 處代表影像 I 在 P 點之前的矩形區域所有像素值的和。如圖 2-18 所示，陰影中的像素之和 $S=A-B-C+D$，僅使用了 3 個加減操作便得出了結果，並且計算時間和矩形區域的大小不相關。

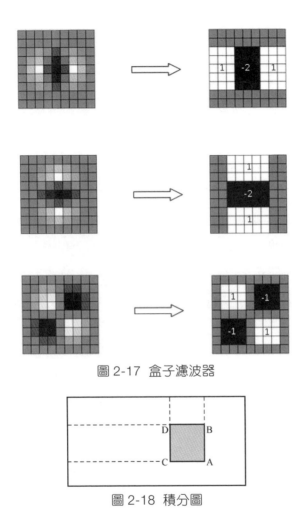

圖 2-17 盒子濾波器

圖 2-18 積分圖

為了使特徵描述符號具有縮放不變性，SURF 與 SIFT 一樣也需要建置連續尺度空間。在 SIFT 演算法中，高斯金字塔中的影像是由其下層影像進行高斯模糊後獲得的，影像間層層依賴，每一層的高斯模糊運算必須等待其下層影像高斯模糊完成後方可進行，這樣就浪費了大量的等待時間。然而 SURF 演算法採用影像不變，盒子濾波器的尺寸逐層變化的策略，進而可以採用平行計算，同時產生金字塔中的各層影像。

SURF 為關鍵點分配主方向的過程也與 SIFT 不同。SURF 演算法首先計算出以關鍵點為中心，以 6s 為半徑的圓形區域內所有像素點的 Harr 小波水平和垂直回應值（Harr 小波水平和垂直方向的濾波器如圖 2-19 所示），然後將兩個值分別乘以對應位置 $\sigma = 2s$ 的高斯核心。s 是關鍵點所在的尺度，它與目前範本的尺寸相關：

$$s = Current\ Filter\ Size \times \frac{Base\ Filter\ Scale}{Base\ Filter\ Size} = Current\ Filter\ Size \times \frac{1.2}{9} \qquad (2\text{-}32)$$

如圖 2-20 所示，使用一個 $\pi/3$ 的扇形視窗計算，並計算該視窗內所有像素 Harr 小波水平和垂直方向回應值之和。然後滑動視窗，找出最大回應值所對應的方向就是主方向。以關鍵點為中心，沿著上一步驟確定的主方向，將其周圍 20s×20s 的區域分為 4×4 的子區域，計算每個子區域內像素點的沿主方向和垂直於主方向的 Harr 小波回應 dx 和 dy，並乘以對應位置 $\sigma = 3.3s$ 的高斯核心。每個子區域統計 $\sum dx$、$\sum dy$ 以及 $\sum |dx|$ 和 $\sum |dy|$ 這 4 個值，這樣在 4×4 子區域就形成了 64 個特徵值，如圖 2-21 所示。和 SIFT 一樣，該特徵描述符號需要進行歸一化處理，以防止光源和比較度的影響。

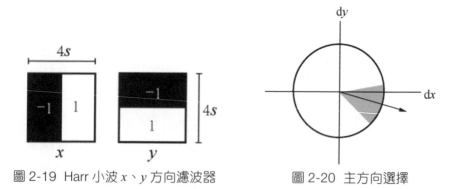

圖 2-19 Harr 小波 *x*、*y* 方向濾波器　　　　圖 2-20 主方向選擇

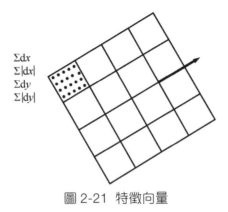

圖 2-21 特徵向量

2.6 本章小結

本章由人類取得和了解一幅影像的基本原理講起，進一步說明了電腦為取得影像而進行的諸如取樣、量化等步驟，以及數位影像儲存的格式、色彩空間等概念，並提供了由這些步驟、概念引出的影像基本操作的實現方式。人類透過影像的特徵來了解它，同樣電腦也是利用特徵來識別和區分影像的。人們將影像的特徵分為全域特徵和局部特徵。全域特徵又由顏色特徵、紋理特徵和形狀特徵組成，文中詳細介紹了這 3 種特徵中典型的演算法和程式實現。局部特徵是影像局部區域的特徵，常常使用某些複雜的數學步驟進行分析和表達，文中對經典的局部特徵演算法 SIFT 和 SURF 進行了介紹。

深度學習影像
特徵分析

3.1 深度學習

1.4 節中曾經提到過目前很多影像搜尋引擎將深度學習演算法引入其中,明顯地改善了影像搜索準確率。2016 年 3 月,世界頂級圍棋棋手李世乭對弈人工智慧棋手 AlphaGo,在 5 場比賽中 AlphaGo 以 4:1 大獲勝。這一人工智慧的偉大勝利將 AlphaGo 所依賴的深度學習演算法的魔力在公眾中做了一個快速的展示。但什麼是深度學習呢?

深度學習是一種多層的神經網路演算法,層的數量代表了它的深度。其實深度學習的演算法早已有之,在 20 世紀 80 年代就已經產生並實際應用到了如今大放異彩的旋積神經網路中。但由於當時條件的限制,訓練一個網路模型常常耗費太多的時間,這一原因也使它並未引起人們的重視,人們普遍認為該方法並不實用。可以說,如今深度學習的成功並不僅是演算法上的成功,而且得益於高速發展的硬體條件和巨量資料的支撐。

3.1.1 神經網路的發展

神經網路演算法起源於 1943 年神經科學家 Warren McCulloch 和數理邏輯學家 Walter Pitts 提出的 MP 神經元模型[1]。神經網路的發展與其他事物的歷史發展過程一樣，都不是一帆風順的。它的發展經歷了兩次低谷，如今以深度學習之名再度復興。

自 19 世紀以來，隨著樹突、軸突、突觸以及髓鞘的相繼發現，人類逐步對神經元有了清晰的認識。神經元又稱為神經細胞，其大小和外觀有很大差異，但都具有細胞體、樹突和軸突，如圖 3-1 所示。神經元透過樹突接受刺激，並將興奮傳入細胞體，一個神經元的軸突末梢突觸和另一個神經元的樹突相連接傳遞興奮，連接部位不同，對神經元的刺激也不同。當神經元接收的所有樹突傳來的電興奮累計而成的膜電位達到一定數值時，神經元才會被啟動進而產生一次脈衝訊號。MP 模型在歸納前人對神經元的生理學研究的基礎上，對其進行數學和網路結構描述，使用二值開關代表單一神經元，使其按不同方式組合能夠完成一定的邏輯運算。如圖 3-2 所示，輸入 $X_1, X_2, X_3, \cdots, X_n$ 與各自對應的權重 $W_1, W_2, W_3, \cdots, W_n$ 加權求和。輸入模擬了神經元受到的各種刺激，權重表示突觸連接不同部位對訊號大小的影響。加權求和後的值模擬了神經元的膜電位，當其值達到規定的設定值 T 時，神經元會被啟動而輸出數值 1，否則輸出為 0。

1　Mcculloch W S. A logical calculus of ideas imminent in nervous activity[J]. Biol Math Biophys, 1943, 5.

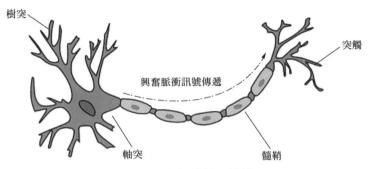

圖 3-1 人類神經元結構

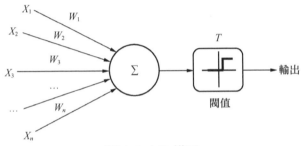

圖 3-2 MP 模型

1957 年，美國康奈爾航空實驗室的 Frank Rosenblatt 在 MP 模型的基礎上，發明了一種叫作「感知器」的神經網路演算法，並在一台 IBM-704 上成功實現[2]。感知器實際上是一種二元線性分類模型，它可以進行簡單的影像識別。感知器的數學表達如下：

$$y_k = \varphi(v_k) = \begin{cases} 1, v_k \geq 0 \\ 0, v_k < 0 \end{cases}, v_k = \sum_{j=1}^{m} W_{kj} X_j + b_k \tag{3-1}$$

其中，W_{kj} 為第 j 個輸入的加權，b_k 為偏置，ϕ 為啟動函數。Rosenblatt 在理論上進一步證明了單層感知器能夠在處理線性可分的模式識別問題

---

2 Rosenblatt F. Perceptron Simulation Experiments[J]. Proceedings of the Ire, 1960, 48(3):301-309.

時收斂,並在此基礎上用實驗證實了感知器具有一定的學習能力。如圖 3-3 所示,透過不斷修正加權 W_{kj} 和偏置 b_k,最後會有一個超平面將樣本空間分為不同的兩類。

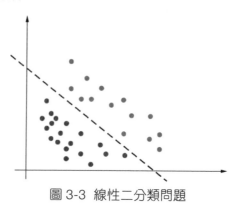

圖 3-3 線性二分類問題

感知器的成功使人們相當大地高估了它的作用。美國海軍曾對感知器寄予厚望,認為它未來可以使計算機具有看、寫、說,甚至複製本身的能力。由此,神經網路開始進入第一次研究熱潮。

然而歷史又總是在潮起潮落間曲折前行。1969 年,人工智慧的先驅 Marvin Minsky 和 Seymour Papert 出版了 *Perceptrons*(《感知器》)一書,書中提出並證明了單層的感知器無法處理不可線性分割的問題,如互斥邏輯,並進一步指出多層感知器也是如此[3]。連簡單的互斥邏輯都不能處理,人們對感知器的熱情也一下降到了冰點,由此神經網路的研究進入了近 20 年的沉寂期。雖然神經網路的研究進入了低谷,但歷史並未停滯。1971 年,蘇聯烏克蘭科學院的 Ivakhnenko 提出了利用 GMDH(Group Method of Data Handling)演算法來訓練一個 8 層的神經網路模型;1974 年,哈佛大學的 Paul Werbos 提出將反向傳播演算法(BP 演

3　Minsky M, Papert S. Perceptrons[J]. American Journal of Psychology, 1969, 84(3):449-452.

算法）的思維應用於神經網路[4]；20 世紀 80 年代初，日本學者福島邦彥提出了可用於解決手寫辨識等模式識別問題的多層神經網路「神經認知機」（Neocognitron）[5]。由於當時的環境，這些極具發展性的事件都未能引起研究者足夠的重視。

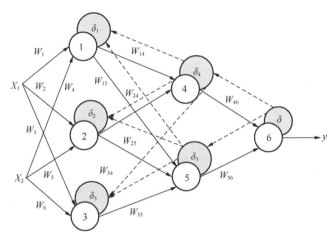

圖 3-4　多層神經網路的 BP 演算法

1986 年，Rumelhart、Hinton 和 Williams 在《自然》雜誌上發表了 *Learning Internal Representation by Backpropagation of Errors* 一文[6]。文中指出，在神經網路中增加一個隱藏層，並使用反向傳播演算法可以解決 Minsky 等人提出的多層神經網路不能解決互斥邏輯的問題。阻礙神經網路發展的魔咒終於被打破了，這也促成了神經網路研究的第二次熱

4　Werbos P. Beyond Regression : New Tools for Prediction and Analysis in the Behavioral Science[J]. Ph.d.dissertation Harvard University, 1974, 29(18):65-78.

5　Fukushima K. Neocognitron: A Self-organizing Neural Network Model for A Mechanism of Pattern Recognition Unaffected by Shift in Position[J]. Biological Cybernetics, 1980, 36(4):193-202.

6　Rumelhart D E, Hinton G E, Williams R J. Learning Internal Representation by Back-propagation of Errors[J]. Nature, 1986, 323(323):533-536.

潮。BP 演算法的基本思維是首先正向計算神經網路的輸出,如圖 3-4 所示,將實際輸出值 y 與目標輸出值 t 相比較得出誤差 δ,將誤差逐層反向傳播,並利用梯度下降演算法對權重進行調整。在經過許多輪反覆運算之後,將最後誤差下降到合理範圍。

1989 年,Yann LeCun 運用旋積神經網路對美國的手寫郵遞區號進行訓練和識別,在獨立樣本測試中達到 5% 的錯誤率,具有很高的實用性。他應用此成果開發的支票自動識別系統,曾經佔據美國近 20% 的市佔率[7]。

1991 年,德國的 SeppHochreiter 指出,當 BP 演算法中成本函數(Cost function)反向傳播時,每經過一層,梯度以相乘的方式覆蓋到前層。梯度在經過許多層反向傳播後會變得極小,趨近於 0,存在梯度消失的問題[8]。剛好在這一時期,以支援向量機(Support Vector Machine,SVM)為代表的統計學習方法因理論嚴謹、效果良好而強勢崛起。由此,神經網路的發展又一次陷入低潮。

3.1.2 深度神經網路的突破

2006 年,Hinton 等人發表了一篇名為 *A Fast Learning Algorithm for Deep Belief Nets* 的論文[9]。該文章描述了他們首先借用統計熱力學中的

7　Y. LeCun, B. Boser, J. S. Denker, D. Henderson, R. E. Howard, W. Hubbard and L. D. Jackel: Backpropagation Applied to Handwritten Zip Code Recognition, Neural Computation, 1(4):541-551, Winter 1989.

8　Sepp Hochreiter. Untersuchungen zu dynamischen neuronalen Netzen. Diploma thesis, TU Munich, 1991.

9　Hinton G E, Osindero S, Teh Y W. A Fast Learning Algorithm for Deep Belief Nets[J]. Neural Computation, 2014, 18(7):1527-1554.

「玻爾茨曼分佈」建置了一種兩層的限制玻爾茨曼機，深信度網路使用幾層覆蓋在一起的限制玻爾茨曼機進行無監督的預訓練，以此來對權重進行初始化，然後使用反向傳播演算法對權重進行微調。這一策略在某種程度上克服了梯度消失的問題。文末，Hinton 進一步天馬行空學習一個更大、更深神經網路的可能性。因此，2006 年也被視為深度學習的起始之年。

2011 年，加拿大蒙特婁大學的 Xavier Glorot 和 Yoshua Bengio 在 *Deep Sparse Rectifier Neural Networks* 的論文中提出一種被稱為「修正線性單元」（rectified linear unit，RELU）的啟動函數[10]。該啟動函數的導數為常數，在誤差反向傳播計算中不存在 sigmoid 等傳統啟動函數所固有的梯度消失問題。這一方法從根本上解決了長期阻礙神經網路發展的梯度消失難題。

2012 年，Hinton 在論文 *Improving neural networks by preventing co-adaptation of feature detectors* 中，提出使用「捨棄」（Dropout）演算法來解決神經網路訓練中的過度擬合問題[11]。Dropout 演算法使用在每次反覆運算訓練神經網路時，隨機刪除每個隱藏層中一定比例神經元的方法來避免過擬合。

在神經網路演算法獲得不斷改進並取得一系列歷史性進步的同期，電腦硬體系統的運算能力也獲得高速發展，各種巨量的資料被標記和整理為專門的訓練資料集，由此神經網路進一步向著「更深」的方向發

10　Glorot X, Bordes A, Bengio Y. Deep Sparse Rectifier Neural Networks[C]// International Conference on Artificial Intelligence and Statistics. 2011:315-323.

11　Hinton G E, Srivastava N, Krizhevsky A, et al. Improving neural networks by preventing co-adaptation of feature detectors[J]. Computer Science, 2012, 3(4):págs. 212-223.

展。現代 CPU 雖然經過 MMX、SSE 系列技術最佳化,但它終究不擅長大規模平行計算,然而 GPU 以單指令流多資料流為基礎的架構,非常適合大量資料平行計算場景。2009 年,史丹佛大學的 Rajat Raina 和 Andrew Ng 在 *Large-scale Deep Unsupervised Learning Using Graphics Processors* 一文中,指出現代影像處理器擁有遠勝於多核心 CPU 的運算能力,它有掀起無監督深度學習方法應用革命的潛力[12]。他們使用 GPU 計算深信度網路和稀疏編碼模型的結果顯示,GPU 比雙核心 CPU 的速度要快 70 倍,並將訓練一個 4 層,多達 1 億個參數的深信度網路模型的時間由幾周降到 1 天。2010 年,Dan Ciresan 等人在名為 *Deep Big Simple Neural Nets Excel on Handwritten Digit Recognition* 的論文中,採用傳統 BP 演算法訓練多層神經網路用於手寫數字識別,在 GPU 上的正向、反向傳播計算要比在雙核心 CPU 上的速度快了 40 倍[13]。在學術界看到 GPU 潛力的同時,產業界也投入重金,不斷研發適用於深度學習且效能更高的 GPU。Nvidia 的 CEO 黃仁勳將人工智慧視為下一個計算浪潮和智慧工業革命,並調整公司戰略,將 Nvidia 由顯示卡廠商轉變為人工智慧計算的裝置公司。在計算硬體進步的同時,各種巨量的訓練資料集也不斷湧現。2007 年,一個名為 "ImageNet" 的圖像資料庫專案開始創立。至 2009 年,它成為了一個擁有 1500 萬幅人工標記影像、22000 個分類的巨大圖像資料集。2016 年,Google 公司發佈了一個包含 800 萬個 YouTube 視訊 URL、4800 個知識圖譜實體的標記視訊資料集。ImageNet 等資料集針對所有研究者開放,相當大地促進了深度學習模型的訓練和技術的發展。

12 Raina R, Madhavan A, Ng A Y. Large-scale Deep Unsupervised Learning Using Graphics Processors[C]// International Conference on Machine Learning. ACM, 2009:873-880.

13 Dan C C, Meier U, Gambardella L M, et al. Deep Big Simple Neural Nets Excel on Handwritten Digit Recognition[J]. Corr, 2010, 22(12):3207-3220.

歷史的每次重大進步並不是偶然發生的，只有當所有的必要條件都具備了，它才會悄然而至。良好的演算法、高性能的計算硬體以及巨量的訓練資料，共同促成了深度學習歷史性的質變。

2010 年，以 ImageNet 圖像資料集為基礎的影像分類大賽 ImageNet Large Scale Visual Recognition Challenge（ILSVRC） 開始每年舉辦一屆。競賽以資料集中的 120 萬張圖片為訓練樣本，將這些影像分為 1000 個不同的類別，然後和影像人工類別標記結果相比較。競賽結果採用 Top-1 和 Top-5 錯誤率標準，也就是對每張影像預測 1 個或 5 個類別，取其中不正確的比例。2012 年，Hinton 和他的兩個研究所學生 Alex Krizhevsky 和 Ilya Sutskever 利用一個 8 層的旋積神經網路，使用了 ReLU 啟動函數和 Dropout 演算法，並採用兩個 GPU 平行計算。他們以 Top-5 錯誤率 17% 的成績在 ILSVRC 中遠超排名第二 Top-5 錯誤率 26.2% 的 SVM 方法。這一結果也標誌著深度神經網路在影像識別領域已大幅領先其他識別技術，堪稱象徵深度神經網路突破的標示性事件。

隨著對神經網路層次的研究不斷加深，人們發現並非層次越多學習能力越強，一個 56 層的深度神經網路識別錯誤率反而高於一個 20 層的神經網路模型。微軟亞洲研究院的何愷明、孫健等人，使用一種被稱為「深度殘餘學習」（Deep Residual Learning）的演算法解決了這一問題，並在 2015 年的 ImageNet 競賽中使用該方法訓練了一個深達 152 層的神經網路模型，使 Top-5 的錯誤率降低到了 3.57%，而一個普通人的錯誤率大概是 5%。這一指標表明深度神經網路在影像識別領域已經超過了人類的水平。

也是在 2012 年，Hinton、鄧力和其他幾位分別代表多倫多大學、微軟、Google、IBM 的研究者聯合發表了一篇的論文 *Deep Neural Networks for Acoustic Modeling in Speech Recognition: The Shared Views of Four Research Groups*[14]。在該論文中，他們提出使用深度神經網路模型 DNN 替代傳統的語音辨識模型 GMM-HMM 中的高斯混合模型（GMM），組成深度神經網路模型與隱馬可夫模型相結合的 DNN-HMM 模型，並將此模型用於語音辨識。在不同語音辨識的基準測試中，DNN-HMM 模型甚至最高可以將 GMM-HMM 模型的錯誤率降低 20% 以上。在 Google 的語音輸入基準測試中，單字錯誤率為 12.3%，有學者將這一成果稱為 20 年來語音辨識領域最大的一次進步。

2013 年，多倫多大學的 Alex Graves 在其論文 *Towards End-to-end Speech Recognition with Recurrent Neural Networks* 中，提出使用 RNN/LSTM 模型來進行語音辨識。他訓練的包含 3 個隱藏層、430 萬參數的 RNN/ LSTM 模型在 TIMIT 基準測試中音位錯誤率達到 17.7%，明顯優於同期其他模型的水平[15]。2015 年，Google 再次使用 RNN/LSTM 技術將 Google 語音輸入的單字錯誤率降到了 8%。同年，百度人工智慧實驗室的 Dario Amodei 等人在 *Deep Speech 2: End-to-End Speech Recognition in English and Mandarin* 中，提出採用一個叫作「封閉循環單元」（GRU）的變種 LSTM 模型進行語音訓練和識別，在 WSJ Eval'92 的基準測試中將單字錯誤率降至 3.1%，在另一個中文基準測試中將錯誤率降至 3.7%，而人類的錯誤率為 5%[16]。

---

14 Hinton G, Deng L, Yu D, et al. Deep Neural Networks for Acoustic Modeling in Speech Recognition: The Shared Views of Four Research Groups[J]. IEEE Signal Processing Magazine, 2012, 29(6):82-97.

15 Graves A, Jaitly N. Towards End-to-end Speech Recognition with Recurrent Neural Networks[C]// International Conference on Machine Learning. 2014:1764-1772.

16 Amodei D, Anubhai R, Battenberg E, et al. Deep Speech 2: End-to-End Speech Recognition in English and Mandarin[J]. Computer Science, 2015.

3.1.3 主要的深度神經網路模型

人們在深度神經網路發展過程中不斷解決其中遇到的各種問題，與此同時，形成了解決各種實際問題的不同神經網路模型。這些模型的種類多達幾十種，有的因為已經有了更好的替代者而不再使用，有的是在某些基本模型類型上做了擴充。目前我們經常使用的深度神經網路模型主要有旋積神經網路（CNN）、遞迴神經網路（RNN）、深信度網路（DBN）、深度自動編碼器（AutoEncoder）和產生對抗網路（GAN）等。

遞迴神經網路實際上包含了兩種神經網路。一種是時間遞迴神經網路，我們通常稱之為循環神經網路（Recurrent Neural Network）；另一種是結構遞迴神經網路（Recursive Neural Network），我們通常稱其為遞迴神經網路，它使用相似的網路結構遞迴形成更加複雜的深度網路。雖然它們都使用相同的英文字首縮寫 RNN，但它們並不具有相同的網路結構。

如圖 3-5 所示，循環神經網路將 t 時刻隱藏層的輸出與 $t+1$ 時刻輸入層的輸入共同輸入至 $t+1$ 時刻的隱藏層，並繼續在時間軸上遞迴下去。在使用時間軸上的反向傳播演算法 BPTT（Back Propagation Through Time）傳遞誤差時同樣會遇到梯度消失的問題。1997 年，Sepp Hochreiter 和 Juergen Schmidhuber 在 *Long Short-term Memory* 一文中，提出可以使用一種叫作長短期記憶單元（LSTM）的技術來解決梯度消失問題[17]。LSTM 借用數位電路中門電路的形式建置了輸入門、遺忘門、輸出門這 3 個控制神經詮譯資訊傳遞的邏輯門，決定某個輸入資訊在一定時間以後是否還需要儲存，何時輸出傳遞到下一神經元，何時捨棄，確保了在必須進行反向傳播時維持固定的誤差。如圖 3-6 所示，門控循環單

---

17 Hochreiter S, Schmidhuber J. Long Short-term Memory.[J]. Neural Computation, 1997, 9(8):1735-1780.

元（Gated Recurrent Unit，GRU）是 LSTM 的一種變形，由 Junyoung Chung 等人在 2014 年提出，它採用一個更新門和一個重置門，功能與 LSTM 類似，速度更快、更易執行，但函數表達力比 LSTM 稍弱[18]。

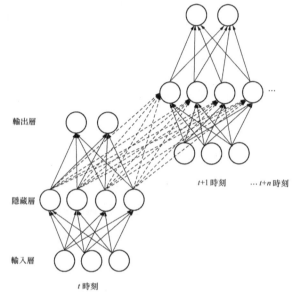

圖 3-5 循環神經網路的 t 和 t+1 時刻的結構圖

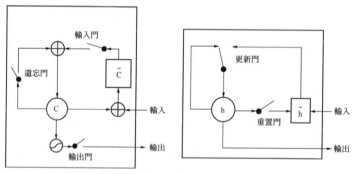

圖 3-6 LSTM 與 GRU

---

18 Chung J, Gulcehre C, Cho K H, et al. Empirical Evaluation of Gated Recurrent Neural Networks on Sequence Modeling[J]. Eprint Arxiv, 2014.

遞迴神經網路（Recursive Neural Network）使用典型的遞迴樹狀結構來組成神經網路，其實是將神經網路用在分析樹上，如圖 3-7 所示。遞迴神經網路正向傳播時依次檢查左孩子、右孩子、根節點，並繼續正向遞迴。當其進行反向傳播誤差時，依次計算根節點、左孩子、右孩子的誤差。由於在遞迴神經網路中，每個節點的加權是相同的，無法表現某些節點的重要性。人們為了解決這一問題，提出了很多遞迴神經網路的變種，例如為每個節點附加一個矩陣的 MV-RNN（Matrix-Vector Recursive Neural Networks），遞迴張量神經網路（Recursive Neural Tensor Network，RNTN）使用以張量為基礎的組合函數來代替 MV-RNN 的附加矩陣，進一步減小了儲存空間需求。

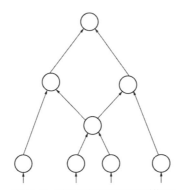

圖 3-7　遞迴神經網路結構圖

產生對抗網路（GAN Generative Adversarial Nets）是一種在 2014 年才剛剛誕生的神經網路模型，由當時還在蒙特婁大學讀博士的 Ian J. Goodfellow 提出[19]。GAN 使用一個產生模型（Generative Model）和一個判別模型（Discriminative Model）。產生模型用於產生更接近自然狀態

19　Goodfellow I J, Pouget-Abadie J, Mirza M, et al. Generative Adversarial Nets[C]// International Conference on Neural Information Processing Systems. MIT Press, 2014:2672-2680.

的資料,而判別模型用於判斷產生模型產生的資料是人工合成的還是自然狀態的。如果達不到要求的相似度,產生模型會繼續產生資料供判別模型判斷。就在產生和判別模型的對抗過程中,產生模型產生了與自然數據相似度很高的人工合成資料。

每種神經網路模型都有其最適用的領域和最擅長解決的問題。深信度網路、旋積神經網路、深度自動編碼器更適合處理影像識別和搜索等影像問題;循環神經網路更適合處理語音辨識、手寫辨識、預測分析等時間序列問題;遞迴神經網路更適合處理語法分析、句法分析、情感分析、詞性標記、語義角色標記等自然語言處理(NLP)問題。在實際工程中,人們還常常將多種神經網路模型聯合使用。將 GAN 和 CNN 結合起來用來產生影像和視訊,例如把模糊影像轉化為清晰影像,去掉視訊或照片中的馬賽克;將 GAN 和 RNN 相結合完成音樂合成以及自然語言的建模。在後續的章節中,我們將詳細介紹和影像搜索相關的旋積神經網路、深信度網路以及深度自動編碼器。

3.2 深度學習應用架構

在深度神經網路的發展過程中,尤其是在近年來深度學習技術逐步走出實驗室進入產業界,實現技術和產業融合發展的處理程序中,研究人員、網際網路公司和開發者研發了許多用於深度學習的應用架構,並開放了架構的原始程式碼,進一步加快了深度學習技術的進步和普及。下面將簡略介紹目前人們常使用的深度學習應用架構。

3.2.1 TensorFlow

TensorFlow 是 Google 公司的 Google Brain 團隊專門針對機器學習和深度神經網路而開發的專用平台。2015 年末，Google 宣佈將 TensorFlow 開放原始碼並公開發佈。TensorFlow 使用資料流程圖的形式來描述計算，資料流程圖中的節點既可以代表數學運算，也可以表示資料登錄的起點、資料輸出的終點或是讀取、寫入持久變數的終點。資料流程圖中的連線表示節點間的輸入、輸出關係，這些資料連線可以傳輸大小能夠動態調整的多維陣列，即張量（Tensor）。張量從資料流程圖中流過的直觀具體也是 "TensorFlow" 得名的緣由。當張量傳入節點以後，節點就被分配到計算裝置上進行非同步、平行的執行。

TensorFlow 並不是只能進行神經網路的計算，只要你能將計算表示成資料流程圖的形式，你就可以使用它進行計算。使用者透過建置圖（Graph）來描述用於驅動計算的運算邏輯，定義新的操作一般只需撰寫 Python 函數，這樣會有很高的程式設計效率。TensorFlow 在 PC、雲、行動裝置上都可執行，能夠實現演算法研究和工程產品的無縫統一。TensorFlow 支援多種程式語言，除了原生支援的 Python 和 C++ 外，還可以透過 SWIG 實現多種語言的呼叫。TensorFlow 支援 GPU，並能夠合理轉換、充分利用硬體資源，將計算單元分配到不同裝置並存執行，具有良好的硬體使用效率。

3.2.2 Torch

Torch 誕生於 2002 年，是一個具有悠久歷史的科學計算程式架構。2015 年，Facebook 開放其在 Torch 之上開發的有關深度學習、可用於電腦視覺和自然語言處理等場景的模組和外掛程式集 fbcunn 原始碼。fbcunn

建置在 Nvidia 發佈的用於深度神經網路的 cuDNN 函數庫之上，大幅提升了 Torch 中原生的 nn 神經網路套件的效能。經 Facebook 改造後的 Torch，採用文字檔設定神經網路模型與程式相分離的模式，支援 GPU 加速計算，具有高度的模組化，提供支援 Android、iOS 行動系統的介面，透過 LuaJIT 連線 C 程式。不過由於 Lua 指令碼語言普及度不高，使得使用 Torch 的開發人員並不太多。

3.2.3 Caffe

Caffe 全稱 Convolutional architecture for fast feature embedding，是當時還在加州大學柏克萊分校攻讀博士學位的賈揚清開發的，並於 2013 年底開放原始碼。Caffe 的設計思維遵循了神經網路由許多層組成的假設，讓使用者可以透過逐層定義的方式組成一個神經網路。在 Caffe 中，Layer 是模型和計算的基本單元，它承擔了神經網路的兩個核心操作 -- 正向傳播和反向傳播。正向傳播接收輸入資料並計算輸出，反向傳播接收關於輸出的梯度來計算相對於輸入的梯度，並反向傳播給它前面的層。只要定義好 layer 的初始化設定（setup）、正向（forward）和反向（backward），就可以將其納入網路中，並最後組成一個由一系列 layer 組成的神經網路（net）。在 Caffe 中，使用 blob 結構來儲存、交換和處理網路中正向、反向傳播的資料和導數。

Caffe 中的模型及其最佳化設定以文字形式存在，並與程式相分離，具有模組化的結構，支援 GPU，擁有快速執行現有成功模型或移轉學習現有模型的能力，因此受到廣大開發者的歡迎。

3.2.4 Theano

Theano 於 2008 年誕生於加拿大蒙特婁大學 LISA 實驗室，是一個整合了 Numpy 的深度學習 Python 軟體套件。Theano 誕生較早，由大量開放原始碼的函數庫組成，為學術界研究人員廣泛使用，許多著名的神經網路函數庫如 Keras、Blocks 均建置在 Theano 之上。

3.2.5 Keras

Keras 是一個基於 TensorFlow、Theano 以及 CNTK 高度封裝的神經網路 API，由 Python 撰寫。Keras 為支援構想的快速實驗而生，它能夠把使用者的想法快速地轉化為結果。Keras 提供一致而簡潔的 API，使用者體驗極好、相當好用，能夠相當大地減少開發者的工作量。在 Keras 中，網路層、損失函數、最佳化器、初始化策略、啟動函數、正規化方法都是獨立的模組，使用者可以靈活地使用它們來建置自己的模型。Keras 具有優良的擴充性，只需要仿照現有的模組撰寫新的類別或函數即可增加新的模組，建立新模組的便利性使得 Keras 更適合於先進的研究工作。Keras 沒有單獨的模型設定檔類型，模型由 Python 程式描述，使其更緊湊、更易偵錯，並提供了擴充的便利性。

3.2.6 DeepLearning4J

DeepLearning4J 是一款以 Java 為基礎的原生深度學習架構，由創業公司 Skymind 於 2014 年 6 月發佈。它是世界上第一個商用等級的深度學習開放原始碼函數庫，主要針對生產環境和商業應用的場景，並可與 Hadoop 和 Spark 大數據系統相整合，隨插即用。DeepLearning4J 對於要在系統中快速整合深度學習功能的開發者尤其受用，包含埃森哲、雪佛蘭、博斯諮詢、IBM 在內的許多業界知名公司都是它的使用者。

由於前面章節的程式都是使用 Java 撰寫的，為保持一致性，深度學習部分的程式採用原生支援 Java 的 DeepLearning4J 函數庫。

3.3 旋積神經網路

3.3.1 旋積

在第 2 章中，我們多次使用濾波運算元進行旋積操作，但旋積究竟什麼呢？下面將詳細對旋積的概念說明。由於電腦處理的都是離散空間的問題，因此下面的討論將僅侷限在離散域上。

一維旋積是一種形如式（3-2）的數學運算：

$$y[n] = f[n] * g[n] = \sum_{i=-\infty}^{\infty} f[i] \times g[n-i] \tag{3-2}$$

其中，* 代表的就是旋積運算子。旋積運算廣泛地應用在許多學科中：在數學中，旋積用來表示一個函數透過另一個函數時，兩個函數有多少重疊的積分；在統計學中，旋積是滑動加權平均；在聲學中，回聲可以表示成源聲與反映各種聲音反射效應函數的旋積；在訊號處理中，一個線性系統的輸出可表示為輸入訊號與系統的脈衝回應的旋積；在影像處理中，旋積可以實現影像模糊、銳化、邊緣檢測等操作。

為了讓讀者更直觀地了解旋積運算，這裡舉一個圖文並茂的實例來說明它的原理。假設 $f(n)=[1,2,3]$，$g(n)=[4,5,6]$，$y(n)=f(n)*g(n)$，那麼旋積結果 $y(n)$ 的長度為 $3+3-1=5$，計算步驟如下：

$y(0)=f(0) \times g(0-0)+f(1) \times g(0-1)+f(2) \times g(0-2)=1 \times 4+2 \times 0+3 \times 0=4$

$y(1)=f(0) \times g(1-0)+f(1) \times g(1-1)+f(2) \times g(1-2)=1 \times 5+2 \times 4+3 \times 0=13$

$y(2)=f(0)\times g(2-0)+f(1)\times g(2-1)+f(2)\times g(2-2)=1\times 6+2\times 5+3\times 4=28$

$y(3)=f(0)\times g(3-0)+f(1)\times g(3-1)+f(2)\times g(3-2)=1\times 0+2\times 6+3\times 5=27$

$y(4)=f(0)\times g(4-0)+f(1)\times g(4-1)+f(2)\times g(4-2)=1\times 0+2\times 0+3\times 6=18$

$y(n)=[4,13,28,27,18]$

旋積公式中的 $g(n-i)$，我們可以換一種方式表示，因 $g(n-i)=g(-i+n)$，故 $g(-i)$ 可以視為以 y 軸為對稱軸將 $g(i)$ 進行翻轉的結果，$g(-i+n)$ 是將翻轉的結果再平移 n。旋積 $y(n)$ 相當於當 $g(-i)$ 不斷平移 n 時與 $f(n)$ 重合部分相乘求和的累積，這也是「旋積」一詞意譯的來源吧。

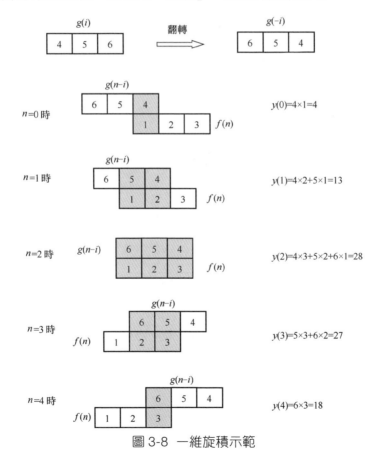

圖 3-8　一維旋積示範

二維離散旋積與一維的情況類似，一個二維旋積就是在二維空間中做水平和垂直方向的一維旋積。二維旋積公式如下：

$$y[m,n] = x[m,n] * h[m,n] = \sum_{j=-\infty}^{\infty} \sum_{i=-\infty}^{\infty} x[i,j] \times h[m-i,n-j] \tag{3-3}$$

在影像處理中，h 常被稱為旋積核心、範本、濾波器。二維旋積的計算等於以下過程：首先將旋積核心 h 分別以 y 軸和 x 軸為對稱軸進行兩次翻轉，將翻轉後的旋積核心 h^* 的中心依次在 3×3 矩陣 x 上移動，計算兩個矩陣重合部分的數值乘積和，並將其作為結果矩陣 y 中旋積核心中心覆蓋位置的數值。在圖 3-9 中，左側是矩陣 x 與翻轉後的 h^* 重合部分求點積的過程，中間是按照公式計算的步驟，右側是每個步驟的結果。由於 x 和 h 未重合時的 x 矩陣對應位置的數值都是零，因此我們在圖 3-9 所示的旋積計算示範步驟第二步中將簡化計算，不再計算數值為零的部分。

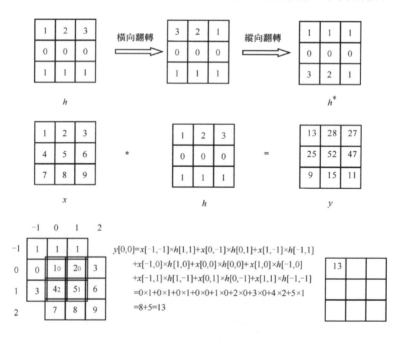

(以下步驟將簡化計算)

圖 3-9 二維旋積示範

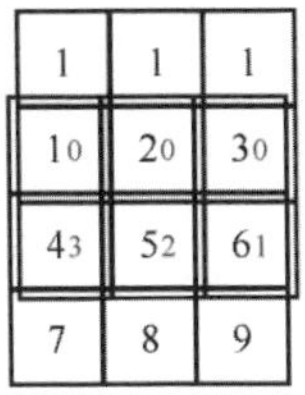

$$y[1,0]=x[0,0]\times h[1,0]+x[1,0]\times h[0,0]+x[2,0]\times h[-1,0]$$
$$+x[0,1]\times h[1,-1]+x[1,1]\times h[0,-1]+x[2,1]\times h[-1,-1]$$
$$=1\times0+2\times0+3\times0+4\times3+5\times2+6\times1$$
$$=12+10+6=28$$

$$y[2,0]=x[1,0]\times h[1,0]+x[2,0]\times h[0,0]$$
$$+x[1,1]\times h[1,-1]+x[2,1]\times h[0,-1]$$
$$=2\times0+3\times0+5\times3+6\times2$$
$$=15+12=27$$

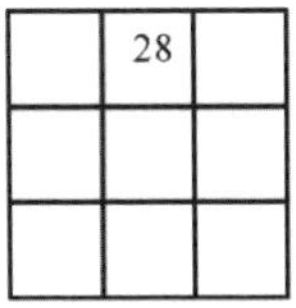

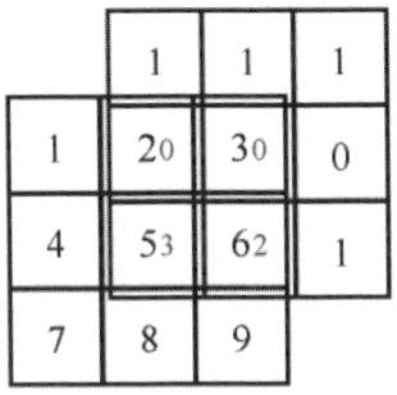

$$y[0,1]=x[0,0]\times h[0,1]+x[1,0]\times h[-1,1]$$
$$+x[0,1]\times h[0,0]+x[1,1]\times h[-1,0]$$
$$+x[0,2]\times h[0,-1]+x[1,2]\times h[-1,-1]$$
$$=1\times1+2\times1+4\times0+5\times0+7\times2+8\times1$$
$$=1+2+14+8=25$$

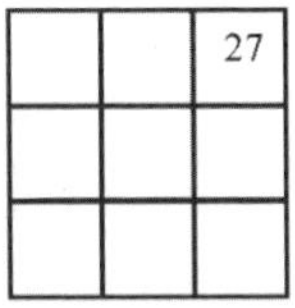

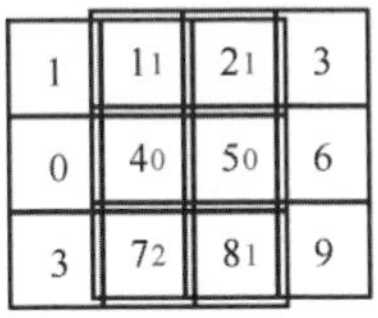

$$y[1,1]=x[0,0]\times h[1,1]+x[1,0]\times h[0,1]+x[2,0]\times h[-1,1]$$
$$+x[0,1]\times h[1,0]+x[1,1]\times h[0,0]+x[2,1]\times h[-1,0]$$
$$+x[0,2]\times h[1,-1]+x[1,2]\times h[0,-1]+x[2,2]\times h[-1,-1]$$
$$=1\times1+2\times1+3\times1+4\times0+5\times0+6\times0+7\times3+8\times2+9\times1$$
$$=1+2+3+21+16+9=52$$

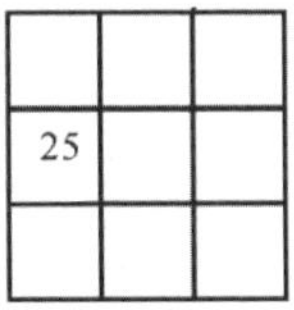

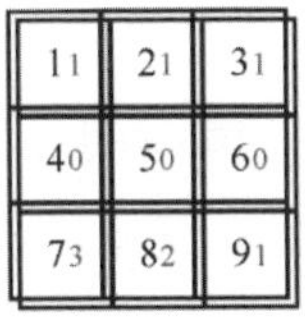

$$y[2,1]=x[1,0]\times h[1,1]+x[2,0]\times h[0,1]$$
$$+x[1,1]\times h[1,0]+x[2,1]\times h[0,0]$$
$$+x[1,2]\times h[1,-1]+x[2,2]\times h[0,-1]$$
$$=2\times1+3\times1+5\times0+6\times0+8\times3+9\times2$$
$$=2+3+24+18=47$$

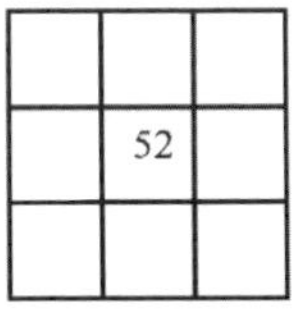

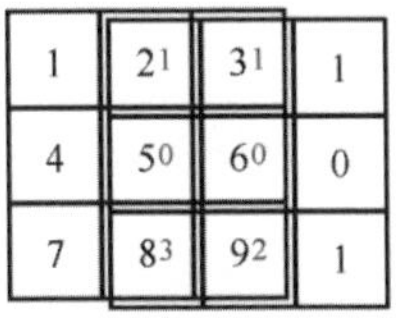

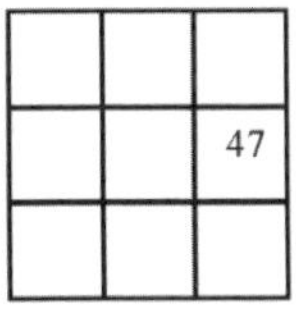

圖 3-9 二維旋積示範（續）

$$y[0,2]=x[0,1]\times h[0,1]+x[1,1]\times h[-1,1]$$
$$+x[0,2]\times h[0,0]+x[1,2]\times h[-1,0]$$
$$=4\times1+5\times1+7\times0+8\times0$$
$$=4+5=9$$

$$y[1,2]=x[0,1]\times h[1,1]+x[1,1]\times h[0,1]+x[2,1]\times h[-1,1]$$
$$+x[0,2]\times h[1,0]+x[1,2]\times h[0,0]+x[2,2]\times h[-1,0]$$
$$=4\times1+5\times1+6\times1+7\times0+8\times0+9\times0$$
$$=4+5+6=15$$

$$y[2,2]=x[1,1]\times h[1,1]+x[2,1]\times h[0,1]$$
$$+x[1,2]\times h[1,0]+x[2,2]\times h[0,0]$$
$$=5\times1+6\times1+8\times0+9\times0$$
$$=5+6=11$$

圖 3-9 二維旋積示範（續）

在進行上面的二維旋積運算示範時，為了方便説明和計算，假設矩陣 x 和旋積核心 h 都為 3×3 的矩陣。在實際的旋積運算中，旋積核心 h 的大小常常要比矩陣 x 小得多。旋積運算過程煩瑣、運算量很大，在實際程式編制中，只為降低二維旋積運算的資源負擔，常常將二維旋積核心分解成兩個一維旋積核心，再進行連續旋積計算。

$$x[m,n]*\begin{bmatrix} A\cdot a & A\cdot b & A\cdot c \\ B\cdot a & B\cdot b & B\cdot c \\ C\cdot a & C\cdot b & C\cdot c \end{bmatrix}=x[m,n]*\left(\begin{bmatrix} A \\ B \\ C \end{bmatrix}\cdot\begin{bmatrix} a & b & c \end{bmatrix}\right)=\left(x[m,n]*\begin{bmatrix} A \\ B \\ C \end{bmatrix}\right)*\begin{bmatrix} a & b & c \end{bmatrix} \tag{3-4}$$

在旋積運算中，對旋積核心的翻轉使得旋積運算具有可交換性。但是在旋積神經網路中並不需要可交換性，所以在旋積神經網路的旋積運算中實際上並未對旋積核心進行翻轉，只是一種點積運算。此外，在旋積神經網路中，輸入影像與旋積核心的旋積運算是以兩者的（0,0）處對齊作為起始的。這些都是需要讀者注意的。

3.3.2 旋積神經網路概述

現在我們已經了解旋積運算的原理、步驟以及它在旋積神經網路中的實現，下面將詳細介紹目前在影像領域大顯神威的旋積神經網路究竟是什麼。

旋積神經網路本質上是一種與多層神經網路 MLP 同樣類型的前饋神經網路。在通常的前饋神經網路中，輸入層接收資料，資料正在傳播，經由許多隱藏層後由輸出層輸出結果。每個隱藏層包含許多神經元，其中每個神經元又與前一層的全部神經元連接，也就是通常所說的全連接。這樣，如果在輸入層輸入一個 300×300 的 RGB 影像，那麼第一個隱藏層的每個神經元就會有 $300 \times 300 \times 3 = 270000$ 個加權。隨著影像的增大，加權的數量會急遽地增加。這一情況不僅造成我們會很快無法找到可以符合的運算能力，而且會使神經網路產生嚴重的過擬合（指神經網路模型與訓練樣本太過比對，以至於無法極佳地實現對新資料的識別和分類）的現象。為了解決這些問題，人們提出局部感受野、參數共用、池化等方法，最後形成我們今天所看到的由許多旋積層、池化層、RELU 層以及全連接層組成的旋積神經網路。下面將一一對這些方法和概念進行詳細解釋。

1 局部感受野與參數共用

為了解決神經元全連接帶來的參數過多問題，人們使用一種稱為局部感受野的方法。如圖 3-10 所示，輸入 16×16 影像的每個像素並未全部與圖中右側隱藏層的神經元相連，而是將輸入影像中一個小區域（3×3 灰色部分）的每個像素與神經元連接，這一個小區域就是隱藏層神經元的局部感受野。局部感受野與神經元的每個連接學習一個加權，並學習一個整體偏置（bias）。然後將局部感受野依次在右、在下移動，每次移動對應一個不同的神經元，這樣就產生了第一個隱藏層，也就是圖 3-10 中 14×14 的隱藏層，我們通常將這個隱藏層叫作「旋積層」。

局部感受野每次移動的像素的距離，稱之為步進值（stride），有時我們會使用不同的步進值，如果每次局部感受野移動兩個像素的距離，那麼 stride=2。局部感受野在每次移動一個步進值的過程中使用的都是同一組加權和偏置，通常稱之為參數共用。3×3 的局部感受野總共有 9 個共用加權，加上一個共用偏置，有 10 個參數，假設使用了 20 個不同的旋積核心來實現特徵對映，那麼總共有 200 個參數。如果採用全連接的神經網路，使用一個相對適中的 20 個神經元的隱藏層，那麼就會有 16×16×20=5120 個加權，加上 20 個偏置，總共 5140 個參數。顯然全連接的參數量是前者的 20 餘倍，隨著影像規模的擴大，這一差距將呈幾何級增加。這樣一來，相比全連接的多層神經網路，使用參數共用策略的旋積神經網路相當大地減少了參數的數量。

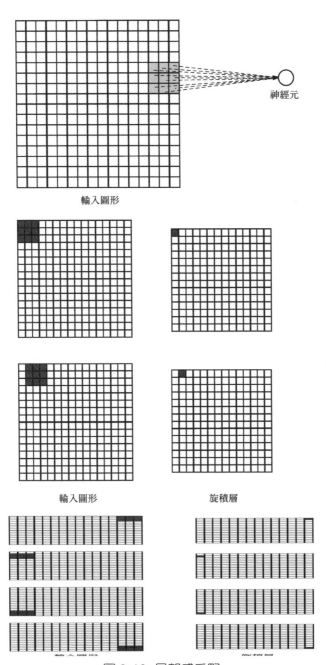

圖 3-10 局部感受野

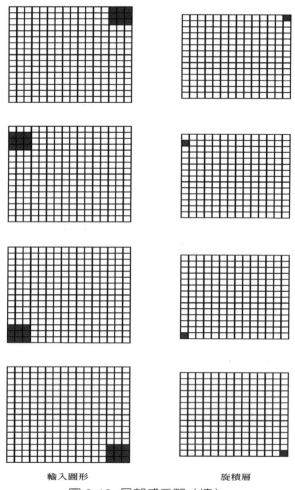

輸入圖形　　　　　　　　　旋積層

圖 3-10 局部感受野（續）

2 零填充

將上面的 16×16 大小的輸入影像進行旋積後，產生 14×14 大小的旋積層。假如照此繼續進行旋積層處理，那麼它縮減的速度常常會超出我們的預期。為了在旋積神經網路的前幾層儘量保留原始輸入資訊，可以採用零填充（zero-padding）的方法擴大輸入影像，如圖 3-11 所示。

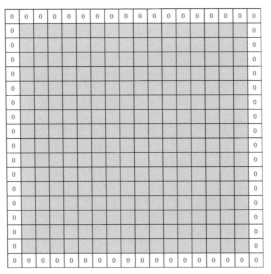

圖 3-11　零填充

3 旋積層

局部感受野中的像素值與各連接加權形成的旋積核心，進行無旋積核心翻轉的旋積運算組成旋積層中的數值。對第（x, y）個隱藏神經元，數值為：

$$o = \sigma\left(b + \sum_{i=0}^{2}\sum_{j=0}^{2} w_{i,j} a_{x+i, y+j} \right) \tag{3-5}$$

σ 是神經元的啟動函數，b 是共用的偏置，$w_{i,j}$ 是 3×3 共用加權矩陣，$a_{x,y}$ 是（x,y）處的輸入像素值。旋積運算的本質實際上是求取相似性，也就是局部感受野內部分輸入影像與旋積核心的相似性。部分輸入影像與旋積核心相似性越高，那麼它們旋積結果的數值越大，使其能夠獲得啟動輸出，反之則不能獲得啟動輸出。在實際的旋積神經網路中，通常使用多個旋積核心來尋找特徵相似性，這樣旋積層中的數值就是局部感受野與每個旋積核心旋積運算結果的總和。

旋積層的輸出大小 O 由輸入的大小 W、局部感受野的大小 F、步進值 S 和填零的數量 P 共同決定,公式如下:

$$O = \frac{W - F + 2P}{S} + 1 \tag{3-6}$$

旋積神經網路通常是由多個旋積層組成的。如圖 3-12 所示,第一個旋積層常常會檢測邊緣、曲線等較低級的特徵,下一個旋積層會學習整合上一個旋積層輸出的特徵,例如第二個旋積層會檢測半圓(曲線和直線的組合)或四邊形(直線的組合)等特徵。隨著網路深度的增加,將獲得更為複雜的特徵,最後我們會檢測到接近檢測物體的特徵,這也是旋積神經網路能夠實現特徵分析和影像分類、識別的基礎。

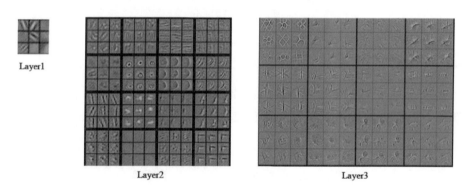

圖 3-12 旋積層的特徵檢測作用

4 RELU 層

在 3.1.2 節中,我們介紹過 RELU 修正線性單元解決了梯度消失和神經網路只能處理線性操作的問題。RELU 層其實就是對所有的輸入內容都應用了 $f(x)=\max(0, x)$ 函數變化的隱藏層。這一層將所有負啟動都變為零,它增加了整個神經網路的非線性特徵。

5 池化層

池化層（Pooling Layer）也叫作下取樣層（Subsampling Layer）。如果我們知道了輸入資料中的特徵，那麼它與其他特徵的相對位置就比它的絕對位置更重要，這樣就可以縮減輸入資料的空間維度。它通常採用平均池化、L2-norm 池化或是最大池化。如圖 3-13 所示，採用了一個 2×2 的篩檢程式和同樣長度的步進值應用到輸入內容上，輸出篩檢程式旋積計算的每個子區域的最大數值。池化層透過池化計算，減少了加權參數的數目，降低計算成本，控制過擬合。

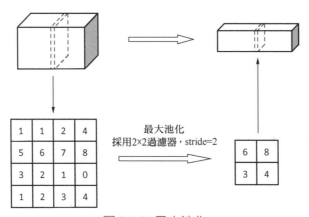

最大池化
採用2×2過濾器，stride=2

圖 3-13　最大池化

6 全連接層

在旋積神經網路中，這一層的每個神經元與前一層的神經元完全連接，故稱為全連接層。一般來說全連接層會輸出一個 N 維向量，N 是神經網路模型能夠分類的數量。例如在手寫阿拉伯數字識別的神經網路模型中會有一個 10 維的向量，向量中的每個數值代表數字 0 ～ 9 的機率。如果模型輸出的結果向量是 [0,0.1,0.1,0,0,0,0.8,0,0.2,0]，那麼它代表檢測的數字是 0、3、4、5、7、9 的機率為 0，是 1 和 2 的機率為 10%，是 6

的機率為 80%，是 8 的機率為 20%。全連接層確定上一層輸出特徵與每個分類的比對度，決定它與誰最為吻合。

７ 超級參數

不同於加權與偏置等能夠從資料訓練中學習到的參數，超級參數並不能從資料中學習到。它需要根據所面對的實際問題，依經驗來事先設定。在深度學習中，我們通常需要根據實際情況設定以下全部或部分超級參數。

(1) 學習率。學習率是最重要的超級參數之一，它表示參數移動到最佳值的速度快慢。如果學習率過大，參數很可能越過最佳值；反之學習率過小，尋找最佳值的過程相當漫長，甚至完全沒有進展。學習率的設定值範圍一般在 0.1 到 1e-6(10^{-6}) 之間，最理想的速率通常取決於實際的資料以及網路架構。我們最初可以在 1e-1(10^{-1})、1e-3(10^{-3})、1e-6(10^{-6}) 三種不同的學習速率間進行嘗試，了解它的大概設定值，然後進一步微調。

(2) 神經網路的層數。

(3) 每個隱藏層中神經元的個數。

(4) 正規化參數。正規化方法透過約束參數的範數，在某種程度上避免了訓練時發生過擬合的情況。

(5) 學習的回合數（Epoch）、反覆運算次數（Iteration）以及微批次數據的大小（Mini-batch size）。在深度學習中，我們通常需要面對極大的訓練資料量。在這種情況下，一次性將訓練資料登錄電腦是不可能的。為了解決這個問題，可以將資料集分為許多小區塊，逐塊傳輸給電腦，在每區塊資料訓練完成後更新一次神經網路的參數。完整地逐塊訓練資料集一次稱為一個回合，完成一個小區塊的訓練叫作一個反覆運算，每個小區區塊資料的大小稱為微批次數據的大小。Epoch、Mini-batch size 都會對模型的最佳化程度和速度產生影響。

(6) 損失函數的選擇。損失函數用以衡量真實值和預測值間的差異程度，損失函數的選擇依據工作的不同而異。對於回歸問題，均方誤差／平方損失函數（L2 損失函數）最為常用；對於分類問題，交叉熵損失函數最為常用。

(7) 加權初始化的方法。加權初始化方法的選擇也會對模型的最佳化程度和速度產生影響。好的初始化方法會加快收斂速度，更易找到最佳解。目前主流的加權初始化方法有高斯分佈初始化、Xavier 初始化等。

(8) 啟動函數的種類。啟動函數能夠為神經網路加入非線性對映能力，使其能夠更進一步地應對複雜問題。啟動函數類型的選擇會直接影響神經網路的收斂速度，對於隱藏層的啟動函數，RELU 啟動函數及其變形一般是比較好的選擇。然而輸出層的啟動函數選擇常常取決於實際的應用，對分類問題而言，通常需要使用 softmax 啟動函數；對於回歸問題而言，恒等啟動函數通常是比較好的選擇。

(9) 梯度最佳化演算法。最常用的方法是隨機梯度下降（SGD）。

(10) 梯度標準化。梯度標準化可以幫助避免梯度在神經網路訓練過程中變得過大或過小。

(11) 參加訓練模型的資料規模。

8 旋積神經網路結構

一個旋積神經網路通常由旋積層、RELU 層、池化層以及全連接層按一定規則組合而成。在一般情況下，可以用下面的模式來描述旋積神經網路的結構：

輸入層→ [[旋積層→ RELU 層]*N →池化層？]*M → [全連接層→ RELU 層]*K →全連接層

其中 * 代表重複，？代表可選的，$N \geq 0$ 且通常 $N \leq 3$，$M \geq 0$，$K \geq 0$ 且通常 $K<3$。舉例來說，當 $N=0$、$M=0$、$K=0$ 時，旋積神經網路會是一個具有輸入層→全連接層結構的線性分類器。當 N、M、K 參數取不同值時，會形成深度不一的旋積神經網路，如：輸入層→旋積層→ RELU 層→全連接層→輸入層→旋積層→ RELU 層→池化層→旋積層→ RELU 層→池化層→全連接層→ RELU 層→全連接層。還有更為複雜的：輸入層→旋積層→ RELU 層→旋積層→ RELU 層→池化層→旋積層→ RELU 層→旋積層→ RELU 層→池化層→旋積層→ RELU 層→旋積層→ RELU 層→池化層→全連接層→ RELU 層→全連接層→ RELU 層→全連接層。

3.3.3 經典旋積神經網路結構

在旋積神經網路的發展過程中，形成了很多經典的旋積神經網路結構，例如 LeNet、AlexNet、ZF Net、GoogLeNet、VGGNet 和 ResNet。

1 LeNet

LeNet 是旋積神經網路第一個成功的商業應用，它由 Yann LeCun 在 20 世紀 90 年代成功開發並運用在手寫支票數字識別和手寫郵遞區號識別領域。如圖 3-14 所示，LeNet-5 由 8 層組成。輸入層接收歸一化為

32×32 大小的手寫數字影像。第一個旋積層 C1 在輸入層上使用 5×5 大小的局部感受野形成 6 個不同的特徵對映，每個特徵對映的輸出大小都為 28×28。C1 層之後是池化層 S2，它是在 C1 層上使用 2×2 大小的感受野經池化操作形成的 6 個 14×14 大小的輸出。緊隨 S2 其後的又是一個旋積層 C3，C3 具有 16 個特徵對映，每個特徵對映連接 S2 上 5×5 大小的局部感受野，輸出大小 10×10。S2 與 C3 的連接比較複雜，連接關係如表 3-1 所示，非全連接的策略把連接的數量控制在合理的規模，並方便不同輸入的特徵分析。S4 是一個在 C3 上使用 2×2 大小的感受野經池化操作形成的 16 個 5×5 大小輸出的池化層。C5 是一個具有 120 個特徵對映的旋積層，每個特徵對映連接 S4 上 5×5 大小的局部感受野，C5 與 S4 全連接。F6 是連接到 C5 的全連接層，具有 84 個神經元。最後的輸出層有 10 個神經元，對應 0 ～ 9 的 10 個數字，採用了歐氏徑向基函數（Eucliean Radial Basis Function，RBF）。RBF 計算輸入與 radial 中心的歐氏距離，RBF 的輸出 y_i 計算公式如下：

$$y_i = \sum_j (x_j - w_{ij})^2 \tag{3-7}$$

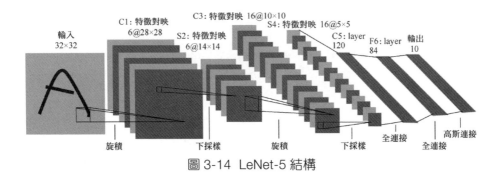

圖 3-14　LeNet-5 結構

RBF 計算了每個輸入與代表性中心的距離，也就是計算了輸入與代表性中心的相似性。

表 3-1 S2 層與 C3 層的連接

	$C3_0$	$C3_1$	$C3_2$	$C3_3$	$C3_4$	$C3_5$	$C3_6$	$C3_7$	$C3_8$	$C3_9$	$C3_{10}$	$C3_{11}$	$C3_{12}$	$C3_{13}$	$C3_{14}$	$C3_{15}$
$S2_0$	√				√	√	√			√	√	√	√		√	√
$S2_1$	√	√				√	√	√			√	√	√	√		√
$S2_2$	√	√	√				√	√	√			√		√	√	√
$S2_3$		√	√	√			√	√	√	√			√		√	√
$S2_4$			√	√	√			√	√	√	√		√	√		√
$S2_5$				√	√	√			√	√	√	√		√	√	√

⧗ 程式 3-1：LeNet 的 DL4J 的實現

```java
package com.ai.deepsearch.deeplearning.models;

import org.deeplearning4j.nn.api.OptimizationAlgorithm;
import org.deeplearning4j.nn.conf.MultiLayerConfiguration;
import org.deeplearning4j.nn.conf.NeuralNetConfiguration;
import org.deeplearning4j.nn.conf.distribution.NormalDistribution;
import org.deeplearning4j.nn.conf.inputs.InputType;
import org.deeplearning4j.nn.conf.layers.ConvolutionLayer;
import org.deeplearning4j.nn.conf.layers.DenseLayer;
import org.deeplearning4j.nn.conf.layers.OutputLayer;
import org.deeplearning4j.nn.conf.layers.SubsamplingLayer;
import org.deeplearning4j.nn.multilayer.MultiLayerNetwork;
import org.deeplearning4j.nn.weights.WeightInit;
import org.nd4j.linalg.activations.Activation;
import org.nd4j.linalg.lossfunctions.LossFunctions;

/**
 * LeNet 模型
 */
public class LeNetModel {
    private int width;
    private int height;
```

```
    private int depth = 3;
    private long seed = 123;
    private int iterations = 90;

    public LeNetModel(int width, int height, int depth, long seed,
                      int iterations) {
        this.width = width;
        this.height = height;
        this.depth = depth;
        this.seed = seed;
        this.iterations = iterations;
    }

    public MultiLayerNetwork initModel() {
        MultiLayerConfiguration.Builder confBuilder = new
NeuralNetConfiguration. Builder()
            // 設定亂數產生器種子
            .seed(seed)
            // 設定最佳化反覆運算次數
            .iterations(iterations)
            // 設定啟動函數
            .activation(Activation.SIGMOID)
            // 設定加權初始化方式
            .weightInit(WeightInit.DISTRIBUTION)
            .dist(new NormalDistribution(0.0, 0.01))
            // 設定學習率
            .learningRate(1e-3)
            .learningRateScoreBasedDecayRate(1e-1)
            .optimizationAlgo(OptimizationAlgorithm.STOCHASTIC_
GRADIENT_DESCENT)
            .list()
            // C1 層，感受野大小 5x5，stride=1
            .layer(0, new ConvolutionLayer.Builder(new int[]{5, 5},
```

```
new int[]{1, 1})
                                .name("C1")
                                .nIn(depth)
                                // C1 層特徵對映數：6
                                .nOut(6)
                                .build())
                                // S2 層，篩檢程式大小 2×2，stride=2
                    .layer(1, new SubsamplingLayer.Builder(SubsamplingLayer.
PoolingType. MAX, new int[]{2, 2}, new int[]{2, 2})
                                .name("S2")
                                .build())
                                // C3 層，感受野大小 5×5，stride=1
                    .layer(2, new ConvolutionLayer.Builder(new int[]{5, 5},
new int[]{1, 1})
                                .name("C3")
                                // C1 層特徵對映數：16
                                .nOut(16)
                                .biasInit(1)
                                .build())
                                // S4 層，篩檢程式大小 2×2，stride=2
                    .layer(3, new SubsamplingLayer.Builder(SubsamplingLayer.
PoolingType. MAX, new int[]{2, 2}, new int[]{2, 2})
                                .name("S4")
                                .build())
                                // C5 層
                    .layer(4, new DenseLayer.Builder()
                                .name("C5")
                                // C1 層特徵對映數：120
                                .nOut(120)
                                .build())
                                // F6 層
                    .layer(5, new DenseLayer.Builder()
                                .name("F6")
```

```
                    // F6 層輸出數：84
                    .nOut(84)
                    .build())
                    // 輸出層
              .layer(6, new OutputLayer.Builder(LossFunctions.
LossFunction.NEGATIVELOGLIKELIHOOD)
                    .name("OUTPUT")
                    // 輸出層輸出 10 個數字 (0 ～ 9) 的機率
                    .nOut(10)
                    .activation(Activation.SOFTMAX)
                    .build())
                    // 反向傳播
              .backprop(true)
                    // 不做預訓練
              .pretrain(false)
                    // 輸入層輸入大小
              .setInputType(InputType.convolutional(height, width,
depth));

        // 根據設定建置網路模型
        MultiLayerNetwork lenetModel = new MultiLayerNetwork(confBuilder.
build());
        lenetModel.init();

        return lenetModel;
    }
}
```

2 AlexNet

AlexNet 是 2012 年由 Hinton 和他的研究所學生 Alex Krizhevsky、Ilya Sutskever 提出的一種深度旋積神經網路結構，我們通常稱之為

AlexNet。AlexNet 在 ILSVRC 中的優異表現也使人們開始重新認識旋積神經網路。如圖 3-15 所示，AlexNet 由輸入層、輸出層和 7 個隱藏層組成。在 7 個隱藏層中，前 5 個是旋積層（有些含有 Max 池化操作），後 2 個是全連接層。最後的輸出層是一個包含 1000 個分類的 softmax，對應 ImageNet 影像函數庫的 1000 個影像類別。輸入層是一幅 224×224 的 RGB 影像，對應 224×224×3 個神經元，第一個旋積層中使用 96 個大小為 11×11×3 的旋積核心，步進值設定為 4，對輸入影像進行旋積操作。由於受到當時單一 GPU 記憶體容量的限制，作者採用兩個 GPU 平行計算的方式，分上下兩部分各處理 96/2=48 個旋積運算。第二個旋積層使用 256（上、下各 128）個大小為 5×5×48 的旋積核心，對第一個旋積層的輸出進行旋積運算，並使用最大池化和局部回應歸一化（LRN）進一步處理旋積後的資料。第三個旋積層使用 384（上、下各 192）個大小為 3×3×256 的旋積核心對第二個旋積層的輸出進行旋積操作，並進行歸一、池化。第四個旋積層使用 384（上、下各 192）個大小為 3×3×192 的旋積核心進行旋積運算，第五個旋積層使用 256（上、下各 128）個大小為 3×3×192 的旋積核心進行旋積。在 5 個旋積層之後是 2 個全連接層，每個全連接層具有 4096（上、下各 2048）個神經元。

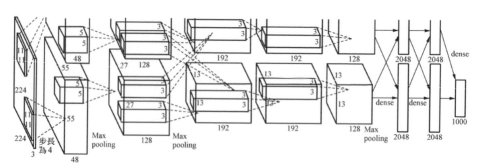

圖 3-15 AlexNet 結構

🖂 程式 3-2：AlexNet 的 DL4J 的實現

```java
package com.ai.deepsearch.deeplearning.models;

import org.deeplearning4j.nn.api.OptimizationAlgorithm;
import org.deeplearning4j.nn.conf. *;
import org.deeplearning4j.nn.conf.distribution.GaussianDistribution;
import org.deeplearning4j.nn.conf.distribution.NormalDistribution;
import org.deeplearning4j.nn.conf.inputs.InputType;
import org.deeplearning4j.nn.conf.layers. *;
import org.deeplearning4j.nn.multilayer.MultiLayerNetwork;
import org.deeplearning4j.nn.weights.WeightInit;
import org.nd4j.linalg.activations.Activation;
import org.nd4j.linalg.lossfunctions.LossFunctions;

/**
 * AlexNet 模型
 */
public class AlexNetModel {
    private int width;
    private int height;
    private int depth = 3;
    private long seed = 123;
    private int iterations = 90;
    private int classfications=1000;

    public AlexNetModel(int width, int height, int depth, long seed,
int iterations, int classfications) {
        this.width = width;
        this.height = height;
        this.depth = depth;
        this.seed = seed;
        this.iterations = iterations;
```

```
       this.classfications = classfications;
    }


    public MultiLayerNetwork initModel() {
        MultiLayerConfiguration.Builder confBuilder=new
NeuralNetConfiguration.Builder()
              // 設定亂數產生器種子
              .seed(seed)
              // 設定加權初始化方式
              .weightInit(WeightInit.DISTRIBUTION)
              .dist(new NormalDistribution(0.0,0.01))
              // 設定啟動函數
              .activation(Activation.RELU)
              // 設定梯度更新器
              .updater(Updater.NESTEROVS)
              // 設定最佳化反覆運算次數
              .iterations(iterations)
              // 設定梯度歸一化策略
              .gradientNormalization(GradientNormalization.
RenormalizeL2PerLayer)
              .optimizationAlgo(OptimizationAlgorithm.STOCHASTIC_
GRADIENT_DESCENT)
              // 設定學習率
              .learningRate(1e-2)
              .biasLearningRate(1e-2*2)
              .learningRateDecayPolicy(LearningRatePolicy.Step)
              // 設定學習率衰退
              .lrPolicyDecayRate(0.1)
              .lrPolicySteps(100000)
              // 使用正規化
              .regularization(true)
              // l2正規化加權係數
              .l2(5*1e-4)
```

```
                    // 僅在梯度更新器設定為 NESTEROVS 時使用
                    .momentum(0.9)
                    // 不執行小量處理輸入
                    .miniBatch(false)
                    .list()
                    //C1層，感受野大小 11x11，stride=4，padding=3
                    .layer(0, new ConvolutionLayer.Builder(new int[]
{11,11}, new int[]{4,4},new int[]{3,3})
                            .name("C1")
                            .biasInit(0)
                            .nIn(depth)
                            .nOut(96)
                            .build())
                    //L2層
                    .layer(1, new LocalResponseNormalization.Builder()
                            .name("L2")
                            .build())
                    //S3層，篩檢程式大小 3x3，stride=2
                    .layer(2, new SubsamplingLayer.Builder(SubsamplingLayer.
PoolingType. MAX, new int[]{3,3},new int[]{2,2})
                            .name("S3")
                            .build())
                    //C4層，感受野大小 5x5，stride=1，padding=2
                    .layer(3,new ConvolutionLayer.Builder(new int[]
{5,5},new int[]{1,1}, new int[]{2,2})
                            .name("C4")
                            .biasInit(1)
                            .nOut(256)
                            .build())
                    //L5層
                    .layer(4, new LocalResponseNormalization.Builder()
                            .name("L5")
                            .build())
```

```
            //S6 層，篩檢程式大小 3x3，stride=2
            .layer(5, new SubsamplingLayer.Builder(SubsamplingLayer.
PoolingType.MAX,new int[]{3,3},new int[]{2,2})
                    .name("S6")
                    .build())
            //C7 層，感受野大小 3x3，stride=1，padding=1
            .layer(6, new ConvolutionLayer.Builder(new int[]
{3,3},new int[]{1,1},new int[]{1,1})
                    .name("C7")
                    .nOut(384)
                    .biasInit(0)
                    .build())
            //C8 層，感受野大小 3x3，stride=1，padding=1
            .layer(7, new ConvolutionLayer.Builder(new int[]
{3,3},new int[]{1,1},new int[]{1,1})
                    .name("C8")
                    .nOut(384)
                    .biasInit(1)
                    .build())
            //C9 層，感受野大小 3x3，stride=1，padding=1
            .layer(8, new ConvolutionLayer.Builder(new int[]
{3,3},new int[]{1,1},new int[]{1,1})
                    .name("C9")
                    .nOut(256)
                    .biasInit(1)
                    .build())
            //S10 層，篩檢程式大小 3x3，stride=2
            .layer(9, new SubsamplingLayer.Builder(SubsamplingLayer.
PoolingType.MAX,new int[]{3,3},new int[]{2,2})
                    .name("S10")
                    .build())
            //F11 層
            .layer(10, new DenseLayer.Builder()
```

```
                        .name("F11")
                        //F11 層輸出數：4096
                        .nOut(4096)
                        .biasInit(1)
                        .dropOut(0.5)
                        .dist(new GaussianDistribution(0,0.005))
                        .build())
                //F12 層
                .layer(11, new DenseLayer.Builder()
                        .name("F12")
                        //F12 層輸出數：4096
                        .nOut(4096)
                        .biasInit(1)
                        .dropOut(0.5)
                        .dist(new GaussianDistribution(0,0.005))
                        .build())
                        // 輸出層，輸出 1000 個分類
                .layer(12, new
OutputLayer.Builder(LossFunctions.LossFunction.NEGATIVELOGLIKELIHOOD)
                        .name("OUTPUT")
                        .nOut(classfications)
                        .activation(Activation.SOFTMAX)
                        .build())
                // 反向傳播
                .backprop(true)
                // 不做預訓練
                .pretrain(false)
                // 輸入層輸入大小
                .setInputType(InputType.convolutional(height, width,
depth));

        // 根據設定建置網路模型
        MultiLayerNetwork alexModel=new MultiLayerNetwork(confBuilder.
```

```
build());
      alexModel.init();


      return alexModel;
   }
}
```

3 GoogLeNet

GoogLeNet 模型是 2014 年 ILSVRC 影像分類工作的冠軍，它的名字之所以寫成這一形式，是為了紀念 LeNet 對旋積神經網路所造成的創新作用。GoogLeNet 具有更深的網路結構，是一個達 22 層的旋積神經網路模型。Google 公司的 Christian Szegedy 等人介紹該模型的論文 *Going deeper with convolution* 中指出：「可以改善深度神經網路效能最常見的辦法，常常是增加深度 -- 神經網路的層數或是增加寬度 -- 每層神經元的數量，然而這種辦法通常更易產生過擬合和導致運算資源的過度需求。解決這一問題的根本辦法是將全連接轉為稀疏連接的結構，而目前的電腦系統對非均勻稀疏資料的計算效率很差。」有沒有一種辦法既能保持網路結構的稀疏性，又能利用密集矩陣的高計算效能呢？文中提出了一種被稱為 "Inception" 的結構來實現這一目的[20]。

Inception 結構的主要思維是使用密整合來近似模擬最佳的局部稀疏結構，如圖 3-16 所示。圖 3-16 的上半部分是 Inception 的初始構想，在多層結構中進行資料相關性統計，將高相關性的區域聚集在一起。接近輸入層的較低層聚集輸入影像的某些區域，逐層聚集，最後獲得在單一區域的大量分群，並透過 1×1、3×3、5×5 尺寸的旋積覆蓋，尺寸越大，分群數量越少。之所以採用大小為 1×1、3×3、5×5 的旋積核

20　Szegedy C, Liu W, Jia Y, et al. Going deeper with convolutions[J]. 2014:1-9.

心,主要是為了方便對齊。此外,可以加一條平行的池化路徑用於提高效率。但是實際使用中,初始版本的 Inception 結構也曝露出一個問題:由於在旋積層頂端的濾波器太多,會帶來過量的運算資源負擔。在加入池化路徑後,這一問題會更加突顯,池化輸入和旋積輸出的融合會導致輸出數量增長得更快。雖然初始版本的 Inception 結構包含了最佳的係數結構,但效率較低。

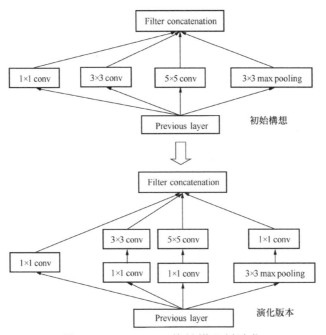

圖 3-16 Inception 的結構及其演化

為了解決這一問題,作者又了改進,提出了演化版本。使用加入維度縮減的方法,在 3×3、5×5 的旋積前用一個 1×1 的旋積來減少計算消耗,並增加修正線性啟動。假設上一層的輸出為 $100 \times 100 \times 128$,經過具有 256 個輸出的 5×5 旋積操作後(stride=1,pad=2),經計算 $(100-5+2 \times 2)/1+1=100$,故輸出為 $100 \times 100 \times 256$,旋積層的參數為

128×5×5×256=819200。如果採用先經過具有 32 個輸出的 1×1 旋積層，再經過具有 256 個輸出的 5×5 旋積層，那麼最後的輸出仍為 100×100×256，但參數量已經減少為 128×1×1×32+32×5×5×256 = 208896，是前一方法參數量的 1/4。

在 GoogLeNet 的網路結構中（如圖 3-17 所示），在較低層使用傳統旋積，而在較高層使用 Inception 結構。在圖 3-17（a）中，為簡化 GoogLeNet 模型表示，將 Inception 結構的一部分用 InceptionN 來表示，如圖 3-17（b）所示。通常採用 Inception 結構的網路模型比沒有採用的網路模型快 2 ～ 3 倍。表 3-2 描述了 GoogLeNet 成功的實例資料，包含 Inception 模組在內的所有旋積都使用了 ReLU。

表 3-2 GoogLeNet 模型成功的實例資料

type	patch size/stride	output size	depth	#1×1	#3×3 reduce	#3×3	#5×5 reduce	#5×5	pool proj	parms	ops
convolution	7×7/2	112×112×64	1							2.7K	34M
max pool	3×3/2	56×56×64	0								
convolution	3×3/1	56×56×192	2		64	192				112K	360M
max pool	3×3/2	28×28×192	0								
inception(3a)		28×28×256	2	64	96	128	16	32	32	159K	128M
inception(3b)		28×28×480	2	128	128	192	32	96	64	380K	304M
max pool	3×3/2	14×14×480	0								
inception(4a)		14×14×512	2	192	96	208	16	48	64	364K	73M
inception(4b)		14×14×512	2	160	112	224	24	64	64	437K	88M
inception(4c)		14×14×512	2	128	128	256	24	64	64	463K	100M
inception(4d)		14×14×528	2	112	144	288	32	64	64	580K	119M
inception(4e)		14×14×832	2	256	160	320	32	128	128	840K	170M
max pool	3×3/2	7×7×832	0								
inception(5a)		7×7×832	2	256	160	320	32	128	128	1072K	54M
inception(5b)		7×7×1024	2	384	192	384	48	128	128	1388K	71M
avg pool	7×7/1	1×1×1024	0								
dropout(40%)		1×1×1024	0								
linear		1×1×1000	1							1000K	1M
softmax		1×1×1000	0								

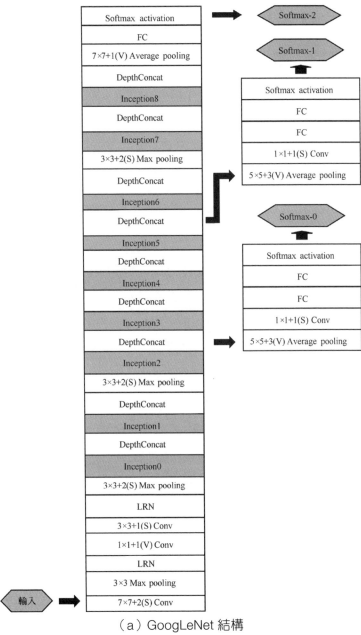

（a）GoogLeNet 結構

圖 3-17

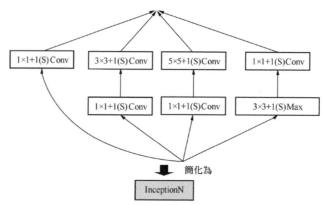

（b）將 Inception 結構的一部分簡化為 InceptionN 表示

圖 3-17（續）

該網路輸入層感受野的大小是 224×224，採用 RGB 影像，並減去平均值。表 3-2 中的 #3×3 reduce 和 #5×5 reduce 分別表示 3×3 和 5×5 旋積前維度縮減層中 1×1 旋積核心的數量，pool proj 表示嵌入最大池化層之後的投影層中 1×1 旋積核心的個數，縮減層和投影層都需要使用 ReLU。該網路中間層產生的特徵非常具有區分性，加了 2 個輔助分類器，見圖 3-17 中的 Softmax-1 和 Softmax-2 部分。

程式 3-3：Inception 的 DL4J 的實現

```java
package com.ai.deepsearch.deeplearning.models;

import org.deeplearning4j.nn.conf.ComputationGraphConfiguration;
import org.deeplearning4j.nn.conf.graph.MergeVertex;
import org.deeplearning4j.nn.conf.layers.ConvolutionLayer;
import org.deeplearning4j.nn.conf.layers.SubsamplingLayer;

/**
 * Inception
 */
```

```
public class Inception {
    public ComputationGraphConfiguration.GraphBuilder
generateInception(ComputationGraphConfiguration.GraphBuilder
graphBuilder, String prefix, int inputSize, int[][] output, String
inputLayer) {
        graphBuilder.addLayer(prefix + "-conv1x1", new ConvolutionLayer.
Builder(new int[]{1, 1}, new int[]{1, 1}, new int[]{0, 0}).
nIn(inputSize).nOut(output[0][0]). biasInit(0.2).build(), inputLayer)
            .addLayer(prefix + "-c3x3reduce", new ConvolutionLayer.
Builder(new int[]{1, 1}, new int[]{1, 1}, new int[]{0, 0}).
nIn(inputSize).nOut(output[1][0]). biasInit(0.2).build(), inputLayer)
            .addLayer(prefix + "-c5x5reduce", new ConvolutionLayer.
Builder(new int[]{1, 1}, new int[]{1, 1}, new int[]{0, 0}).
nIn(inputSize).nOut(output[2][0]).biasInit(0.2).build(), inputLayer)
            .addLayer(prefix + "-maxpool", new SubsamplingLayer.
Builder(new int[]{3, 3}, new int[]{1, 1}, new int[]{1, 1}).build(),
inputLayer)
            .addLayer(prefix + "-conv3x3", new ConvolutionLayer.
Builder(new int[]{3, 3}, new int[]{1, 1}, new int[]{1, 1}).nIn(output[1]
[0]).nOut(output[1][1]).biasInit(0.2).build(), prefix + "-c3x3reduce")
            .addLayer(prefix + "-conv5x5", new ConvolutionLayer.
Builder(new int[]{5, 5}, new int[]{1, 1}, new int[]{2, 2}).nIn(output[2]
[0]).nOut(output[2][1]).biasInit(0.2).build(), prefix + "-c5x5reduce")
            .addLayer(prefix + "-poolproj", new ConvolutionLayer.
Builder(new int
[]{1, 1}, new int[]{1, 1}, new int[]{0, 0}).nIn(inputSize).
nOut(output[3][0]).biasInit(0.2).build(), prefix + "-maxpool")
            .addVertex(prefix + "-depthconcat", new MergeVertex(),
prefix + "-conv1x1", prefix + "-conv3x3" + prefix + "-conv5x5", prefix
+ "-poolproj");

        return graphBuilder;

    }

}
```

⧗ 程式 3-4：GoogLeNet 的 DL4J 的實現

```java
package com.ai.deepsearch.deeplearning.models;

import org.deeplearning4j.nn.api.OptimizationAlgorithm;
import org.deeplearning4j.nn.conf.ComputationGraphConfiguration;
import org.deeplearning4j.nn.conf.LearningRatePolicy;
import org.deeplearning4j.nn.conf.NeuralNetConfiguration;
import org.deeplearning4j.nn.conf.Updater;
import org.deeplearning4j.nn.conf.layers. *;
import org.deeplearning4j.nn.graph.ComputationGraph;
import org.deeplearning4j.nn.weights.WeightInit;
import org.nd4j.linalg.activations.Activation;
import org.nd4j.linalg.lossfunctions.LossFunctions;

/**
 *  GoogLeNet 模型
 */
public class GoogLeNetModel {
    private int width;
    private int height;
    private int depth = 3;
    private long seed = 123;
    private int iterations = 90;
    private int classfications = 1000;

    public GoogLeNetModel(int width, int height, int depth, long seed,
                          int iterations, int classfications) {
        this.width = width;
        this.height = height;
        this.depth = depth;
        this.seed = seed;
        this.iterations = iterations;
```

```
        this.classfications = classfications;
    }


    public ComputationGraph initModel() {
        ComputationGraphConfiguration.GraphBuilder graphBuilder = new
NeuralNetConfiguration.Builder()
                // 設定亂數產生器種子
        .seed(seed)
                // 設定加權初始化方式
                .weightInit(WeightInit.XAVIER)
                // 設定啟動函數
                .activation(Activation.RELU)
                // 設定最佳化反覆運算次數
                .iterations(iterations)
                .optimizationAlgo(
                    OptimizationAlgorithm.STOCHASTIC_GRADIENT_DESCENT)
                // 設定學習率
                .learningRate(1e-2).biasLearningRate(2 * 1e-2)
                .learningRateDecayPolicy(LearningRatePolicy.Step)
                .lrPolicyDecayRate(0.96)
                .lrPolicySteps(320000)
                // 使用正規化
                .regularization(true)
                // l2 正規化加權係數
                .l2(2e-4)
                // 設定梯度更新器
                .updater(Updater.NESTEROVS)
                // 僅在梯度更新器設定為 NESTEROVS 時使用
                .momentum(0.9)
                .graphBuilder();
        // 設定輸入層
        graphBuilder.addInputs("INPUT")
```

```
                .addLayer("C1", new ConvolutionLayer.Builder(new
int[]{7,7},new int[]{2,2}, new int[]{3,3}).nIn(depth).nOut(64).
biasInit(0.2).build(),"INPUT")
                .addLayer("S2", new SubsamplingLayer.Builder(new int[]
{3,3}, new int[]{2,2}, new int[]{0,0}).build(), "C1")
                .addLayer("L3", new LocalResponseNormalization.
Builder(5, 1e-4, 0.75).build(), "S2")
                .addLayer("C4", new ConvolutionLayer.Builder(new int[]
{1,1},new int[]{1,1}, new int[]{0,0}).nIn(64).nOut(64).biasInit(0.2).
build(), "L3")
                .addLayer("C5", new ConvolutionLayer.Builder(new int[]
{3,3},new int[]{1,1}, new int[]{1,1}).nIn(64).nOut(192).biasInit(0.2).
build(), "C4")
                .addLayer("L6", new LocalResponseNormalization.
Builder(5, 1e-4, 0.75).build(), "C5")
                .addLayer("S7", new SubsamplingLayer.Builder(new int[]
{3,3}, new int[]{2,2}, new int[]{0,0}).build(), "L6");
        // Inception 物件
        Inception inception=new Inception();
        inception.generateInception(graphBuilder, "3a", 192, new int[]
[]{{64},{96,128},{16,32},{32}}, "S7");
        inception.generateInception(graphBuilder, "3b", 256, new int[]
[]{{128},{128,192},{32,96},{64}}, "3a-depthconcat");
        graphBuilder.addLayer("S8", new SubsamplingLayer.Builder(new
int[]{3,3}, new int[]{2,2}, new int[]{0,0}).build(), "3b-depthconcat");
        inception.generateInception(graphBuilder, "4a", 480, new int[]
[]{{192},{96,208},{16,48},{64}}, "3b-depthconcat");
        inception.generateInception(graphBuilder, "4b", 512,new int[][]
{{160},{112,224},{24,64},{64}} , "4a-depthconcat");
        inception.generateInception(graphBuilder, "4c", 512,new int[][]
{{128},{128,256},{24,64},{64}} , "4b-depthconcat");
        inception.generateInception(graphBuilder, "4d", 512,new int[][]
```

```
{{112},{144,288},{32,64},{64}} , "4c-depthconcat");
        inception.generateInception(graphBuilder, "4e", 528,new
int[][]{{256},{160,320},{32,128},{128}} , "4d-depthconcat");
        graphBuilder.addLayer("S8", new SubsamplingLayer.Builder(new
int[]{3,3}, new int[]{2,2}, new int[]{0,0}).build(), "4e-depthconcat");
        inception.generateInception(graphBuilder, "5a", 832,new int[][]
{{256},{160,320},{32,128},{128}} , "S8");
        inception.generateInception(graphBuilder, "5b", 832,new int[][]
{{384},{192,384},{48,128},{128}} , "5a-depthconcat");
        graphBuilder.addLayer("S9", new SubsamplingLayer.
Builder(SubsamplingLayer.PoolingType.AVG,new int[]{7,7}, new int[]
{2,2}, new int[]{0,0}).build(), "5b-depthconcat")
                .addLayer("F10", new DenseLayer.Builder().nIn(1024).
nOut(1024).dropOut(0.4).build(), "S9")
                .addLayer("OUTPUT", new OutputLayer.Builder
(LossFunctions.LossFunction.NEGATIVELOGLIKELIHOOD).nIn(1024).
nOut(classfications).activation(Activation.SOFTMAX).build(), "F10")
                // 設定輸出層
                .setOutputs("OUTPUT")
                // 反向傳播
                .backprop(true)
                // 不做預訓練
                .pretrain(false);

        // 根據設定建置網路模型
        ComputationGraphConfiguration conf=graphBuilder.build();
        ComputationGraph googlenetModel=new ComputationGraph(conf);
        googlenetModel.init();

        return googlenetModel;
    }
}
```

4 VGGNet

VGGNet 是由牛津大學的視覺幾何團隊提出的網路模型結構，在 2014
年的 ILSVRC 中取得分類工作第二名、定位工作第一名的成績。和
GoogLeNet 一，VGGNet 也利用更小的旋積核心（3×3 大小的旋積核
心）來實現網路向更深方向發展的策略。與 AlexNet 在第一個旋積層使
用 11×11 大小的旋積核心不同，VGGNet 使用多個連續的 3×3 旋積操
作達到相同的目的。兩個連續的 3×3 旋積相當於一個 5×5 的旋積，
3 個相當於一個 7×7 的旋積。使用連續的 3×3 旋積，而不使用大尺寸
的旋積操作，具有兩個突出的優點：一是每個 3×3 旋積層後都會使用
ReLU，這樣就包含了多個 ReLU，使決策函數更具判別性；二是進一步
減少了參數，和在說明 Inception 結構中描述的一，多個連續的小尺寸旋
積比一個大尺寸的旋積使用更少的參數。

如表 3-3 所示，從 A 到 E，網路的深度逐漸加深，更多的層被加入，加
入的層用粗體表示。旋積層的參數用 conv< 感受野大小 >-< 通道數量 >
的形式表示，為了簡潔，表中未表現 ReLU 啟動函數。A 具有 11 層，包
含 8 個旋積層和 3 個全連接層。E 具有 19 層，包含 16 個旋積層和 3 個
全連接層。輸入層接收 224×224 大小的 RGB 輸入影像，並減去像素平
均值。旋積層通道數從 64 到 512，每經過一個最大池化層，數量擴大一
倍。前兩個全連接層都具有 4096 個通道，第三個全連接層具有 1000 個
通道，用於分類。最後一層是 Softmax 層。

表 3-3　VGGNet 模型網路結構

VGGNet 設定					
A	A-LRN	B	C	D	E
11 層	11 層	13 層	16 層	16 層	19 層
輸入層（224×224 RGB 影像）					
conv3-64	conv3-64 LRN	conv3-64 conv3-64	conv3-64 conv3-64	conv3-64 conv3-64	conv3-64 conv3-64
最大池化層					
conv3-128	conv3-128	conv3-128 conv3-128	conv3-128 conv3-128	conv3-128 conv3-128	conv3-128 conv3-128
最大池化層					
conv3-256 conv3-256	conv3-256 conv3-256	conv3-256 conv3-256	conv3-256 conv3-256 conv1-256	conv3-256 conv3-256 conv3-256	conv3-256 conv3-256 conv3-256 conv3-256
最大池化層					
conv3-512 conv3-512	conv3-512 conv3-512	conv3-512 conv3-512	conv3-512 conv3-512 conv1-512	conv3-512 conv3-512 conv3-512	conv3-512 conv3-512 conv3-512 conv3-512
最大池化層					
conv3-512 conv3-512	conv3-512 conv3-512	conv3-512 conv3-512	conv3-512 conv3-512 conv1-512	conv3-512 conv3-512 conv3-512	conv3-512 conv3-512 conv3-512 conv3-512
最大池化層					
全連接層 -4096					
全連接層 -4096					
全連接層 -1000					
Softmax 層					

⏳ 程式 3-5：VGGNet 的 DL4J 的實現

```java
package com.ai.deepsearch.deeplearning.models;

import org.deeplearning4j.nn.api.OptimizationAlgorithm;
import org.deeplearning4j.nn.conf.GradientNormalization;
import org.deeplearning4j.nn.conf.MultiLayerConfiguration;
import org.deeplearning4j.nn.conf.NeuralNetConfiguration;
import org.deeplearning4j.nn.conf.Updater;
import org.deeplearning4j.nn.conf.distribution.NormalDistribution;
import org.deeplearning4j.nn.conf.inputs.InputType;
import org.deeplearning4j.nn.conf.layers.ConvolutionLayer;
import org.deeplearning4j.nn.conf.layers.DenseLayer;
import org.deeplearning4j.nn.conf.layers.OutputLayer;
import org.deeplearning4j.nn.conf.layers.SubsamplingLayer;
import org.deeplearning4j.nn.multilayer.MultiLayerNetwork;
import org.deeplearning4j.nn.weights.WeightInit;
import org.nd4j.linalg.activations.Activation;
import org.nd4j.linalg.lossfunctions.LossFunctions;

/**
 *  VGGNet 模型
 */
public class VGGNetModel {
    private int width;
    private int height;
    private int depth = 3;
    private long seed = 123;
    private int iterations = 370;
    private int classfications=1000;

    public VGGNetModel(int width, int height, int depth, long seed, int
iterations,int classfications) {
```

```java
        this.width = width;
        this.height = height;
        this.depth = depth;
        this.seed = seed;
        this.iterations = iterations;
        this.classfications = classfications;
    }


    public MultiLayerNetwork initModel() {
        MultiLayerConfiguration.Builder confBuilder=new
NeuralNetConfiguration.Builder()
                // 設定亂數產生器種子
                .seed(seed)
                // 設定加權初始化方式
                .weightInit(WeightInit.DISTRIBUTION)
                .dist(new NormalDistribution(0.0,0.01))
                // 設定啟動函數
                .activation(Activation.RELU)
                // 設定梯度更新器
                .updater(Updater.NESTEROVS)
                // 設定最佳化反覆運算次數
                .iterations(iterations)
                // 設定梯度歸一化策略
                .gradientNormalization(GradientNormalization.
RenormalizeL2PerLayer)
                .optimizationAlgo(OptimizationAlgorithm.STOCHASTIC_
GRADIENT_DESCENT)
                // 設定學習率
                .learningRate(1e-1)
                .learningRateScoreBasedDecayRate(1e-1)
                // 使用正規化
                .regularization(true)
```

```
                // 12 正規化加權係數
                .l2(5*1e-4)
                // 僅在梯度更新器設定為 NESTEROVS 時使用
                .momentum(0.9)
                .list()
                //C1 層，感受野大小 3x3，stride=1，padding=1
                .layer(0, new ConvolutionLayer.Builder(new int[]{3,3},
new int[]{1,1},new int[]{1,1})
                        .name("C1")
                        .nIn(depth)
                        .nOut(64)
                        .build())
                //S2 層，篩檢程式大小 2x2
                .layer(1, new SubsamplingLayer.Builder(SubsamplingLayer.
PoolingType.MAX, new int[]{2,2})
                        .name("S2")
                        .build())
                //C3 層，感受野大小 3x3，stride=1，padding=1
                .layer(2,new ConvolutionLayer.Builder(new int[]
{3,3},new int[]{1,1},new int[]{1,1})
                        .name("C3")
                        .nOut(128)
                        .build())
                //S4 層，篩檢程式大小 2x2
                .layer(3, new SubsamplingLayer.Builder(SubsamplingLayer.
PoolingType.MAX, new int[]{2,2})
                        .name("S4")
                        .build())
                //C5 層，感受野大小 3x3，stride=1，padding=1
                .layer(4,new ConvolutionLayer.Builder(new int[]
{3,3},new int[]{1,1},new int[]{1,1})
                        .name("C5")
```

```
                            .nOut(256)
                            .build())
                //C6層，感受野大小 3x3，stride=1，padding=1
                .layer(5,new ConvolutionLayer.Builder(new int[]
{3,3},new int[]{1,1}, new int[]{1,1})
                            .name("C5")
                            .nOut(256)
                            .build())
                //S7層，篩檢程式大小 2x2
                .layer(6, new SubsamplingLayer.Builder(SubsamplingLayer.
PoolingType. MAX, new int[]{2,2})
                            .name("S7")
                            .build())
                //C8層，感受野大小 3x3，stride=1，padding=1
                .layer(7,new ConvolutionLayer.Builder(new int[]{3,3},
new int[]{1,1}, new int[]{1,1})
                            .name("C8")
                            .nOut(512)
                            .build())
                //C9層，感受野大小 3x3，stride=1，padding=1
                .layer(8,new ConvolutionLayer.Builder(new int[]{3,3},
new int[]{1,1}, new int[]{1,1})
                            .name("C9")
                            .nOut(512)
                            .build())
                //S10層，篩檢程式大小 2x2
                .layer(9, new SubsamplingLayer.Builder(SubsamplingLayer.
PoolingType. MAX, new int[]{2,2})
                            .name("S10")
                            .build())
                //C11層，感受野大小 3x3，stride=1，padding=1
                .layer(10,new ConvolutionLayer.Builder(new int[]{3,3},
```

```
new int[]{1,1},new int[]{1,1})
                        .name("C11")
                        .nOut(512)
                        .build())
                //C12層，感受野大小 3x3，stride=1，padding=1
                .layer(11,new ConvolutionLayer.Builder(new int[]{3,3},
new int[]{1,1},new int[]{1,1})
                        .name("C12")
                        .nOut(512)
                        .build())
                //S13層，篩檢程式大小 2x2
                .layer(12, new SubsamplingLayer.Builder
(SubsamplingLayer.PoolingType.MAX, new int[]{2,2})
                        .name("S13")
                        .build())
                //F14層
                .layer(13, new DenseLayer.Builder()
                        .name("F14")
                        //F14層輸出數：4096
                        .nOut(4096)
                        .dropOut(0.5)
                        .build())
                //F15層
                .layer(14, new DenseLayer.Builder()
                        .name("F15")
                        //F15層輸出數：4096
                        .nOut(4096)
                        .dropOut(0.5)
                        .build())
                // 輸出層，輸出 1000 個分類
                .layer(15, new OutputLayer.Builder(LossFunctions.
LossFunction.NEGATIV ELOGLIKELIHOOD)
```

```
                            .name("OUTPUT")
                            .nOut(classfications)
                            .activation(Activation.SOFTMAX)
                            .build())
            // 反向傳播
            .backprop(true)
            // 不做預訓練
            .pretrain(false)
            // 輸入層輸入大小
            .setInputType(InputType.convolutional(height, width,
depth));

        // 根據設定建置網路模型
        MultiLayerNetwork vggModel=new MultiLayerNetwork(confBuilder.
build());
        vggModel.init();

        return vggModel;
    }
}
```

3.3.4 使用旋積神經網路分析影像特徵

在 2.3 節中，我們曾經提到影像特徵具有穩健性的特點，也就是説在影像經過平移、縮放、旋轉等操作後，影像特徵仍能穩定的代表該影像。那麼旋積神經網路分析的特徵是否也具有這些特點呢？

由於權重共用的機制和池化層的作用，旋積神經網路在一定範圍內具有平移、縮放、旋轉的不變性。權重共用使其具有一定空間範圍內的特徵檢測能力，池化操作進一步降低了它對局部移動和形變的敏感性。例如讀者可以觀察圖 3-18，它抽象地描述了池化層的作用。可以想像一下，圖中的灰色小區塊無論在大的矩形框內怎樣移動，它的池化結果都是不會變化的。然而旋積神經網路分析特徵並不具有大空間範圍內的平移不變性，以及良好的縮放和旋轉不變性，它們需要旋積神經網路透過學習有關的資料而獲得。通常用來訓練模型的大類型資料集 ImageNet 中存在大量不同角度拍攝、焦距不一、物體位置不同、曝光度不同的同一事物的照片，如圖 3-19 所示。這也使大量在其上預訓練的模型分析的特徵具有了更大範圍內的平移、縮放、旋轉不變性。

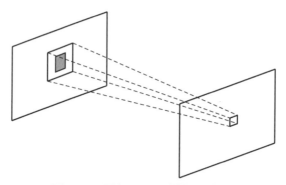

圖 3-18 局部平移不變性示意圖

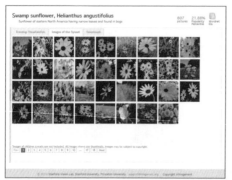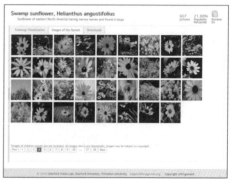

圖 3-19 ImageNet 資料集中的實例

如果我們想要分析效能更好、範圍更大的不變性特徵，就需要對原始影像採用多方向、多尺寸的平移，不同角度的旋轉，不相較去年例的縮放

等一系列資料增強方法擴充資料集，進而獲得更具穩健性的不變特徵，
程式如下。

⧗ **程式 3-6：資料集擴充**

```java
package com.ai.deepsearch.deeplearning.datasets;

import com.ai.deepsearch.deeplearning.models.AlexNetModel;
import com.ai.deepsearch.deeplearning.models.LeNetModel;
import com.ai.deepsearch.deeplearning.models.VGGNetModel;
import org.datavec.api.io.labels.ParentPathLabelGenerator;
import org.datavec.api.split.FileSplit;
import org.datavec.api.split.InputSplit;
import org.datavec.image.loader.BaseImageLoader;
import org.datavec.image.recordreader.ImageRecordReader;
import org.datavec.image.transform. *;
import org.deeplearning4j.datasets.datavec.RecordReaderDataSetIterator;
import org.deeplearning4j.datasets.iterator.MultipleEpochsIterator;
import org.deeplearning4j.nn.multilayer.MultiLayerNetwork;
import org.nd4j.linalg.dataset.api.iterator.DataSetIterator;
import org.nd4j.linalg.dataset.api.preprocessor.DataNormalization;
import org.nd4j.linalg.dataset.api.preprocessor.ImagePreProcessingScaler;

import java.io.File;
import java.io.IOException;
import java.nio.file.Paths;
import java.util.Random;

/**
 * 擴充資料集
 */
public class DataAugmentation {
    private MultiLayerNetwork loadModel(String modelType, int width,
```

```java
int heigth, int channels) {
        MultiLayerNetwork network = null;
        switch (modelType) {
            case "LeNet":
                LeNetModel leNetModel = new LeNetModel(width, heigth,
channels, 123, 90);
                network = leNetModel.initModel();
                break;
            case "AlexNet":
                AlexNetModel alexNetModel = new AlexNetModel(width,
heigth, channels, 123, 90, 1000);
                network = alexNetModel.initModel();
                break;
            case "VGGNet":
                VGGNetModel vggNetModel = new VGGNetModel(width,
heigth, channels, 123, 370, 1000);
                network = vggNetModel.initModel();
                break;
        }
        return network;
    }

    private ImageTransform[] getMultiTransforms() {
        // 以隨機方式旋轉影像
        ImageTransform rotateTransform = new RotateImageTransform(new
Random(123), 360);
        // 以隨機方式縮放影像
        ImageTransform scaleTransform = new ScaleImageTransform(new
Random(123), 1);
        // 以隨機方式翻轉影像
        ImageTransform flipTransform = new FlipImageTransform(new
Random(123));
```

```
        return new ImageTransform[]{rotateTransform, scaleTransform,
flipTransform};
    }

    // 利用擴充資料集進行訓練
    private void augmentTrain(int width, int height, int channels, int
labelNums, int miniBatchSize, int epochs) throws IOException {
        ImageTransform[] trans = getMultiTransforms();
        ParentPathLabelGenerator labelGen = new
ParentPathLabelGenerator();
        File trainDir = Paths.get("resource/datasets/train").toFile();
        InputSplit trainData = new FileSplit(trainDir, BaseImageLoader.
ALLOWED_FORMATS, new Random());
        ImageRecordReader trainReader = new ImageRecordReader(height,
width, channels, labelGen);
        DataNormalization scaler = new ImagePreProcessingScaler(0, 1);
        DataSetIterator dataIter;
        MultipleEpochsIterator trainIter;
        for (ImageTransform tran : trans) {
            trainReader.initialize(trainData, tran);
            dataIter = new RecordReaderDataSetIterator(trainReader,
miniBatchSize, 1, labelNums);
            scaler.fit(dataIter);
            dataIter.setPreProcessor(scaler);
            MultiLayerNetwork network = loadModel("VGGNet", width,
height, channels);
            trainIter = new MultipleEpochsIterator(epochs, dataIter);
            network.fit(trainIter);
        }
    }
}
```

目前利用旋積神經網路分析的特徵通常都是使用在 ImageNet 資料集上進行預訓練的模型，諸如 AlexNet、VGGNet 等在某個全連接層或較後面的旋積層上所形成的特徵。在旋積神經網路中，處於較前面的旋積層識別物體的邊緣特徵，後面的其他旋積層逐層對前面的特徵進行抽象，由邊緣特徵到角和輪廓，再到物體的某個部分，而全連接層分析的特徵更為考慮全域，是高度提純和壓縮的特徵。目前在影像檢索領域，我們通常採用更有效的以 ImageNet 資料集預訓練為基礎的 AlexNet 和 VGGNet 的某個全連接層來分析影像特徵。見以下程式，利用 16 層的預訓練 VGGNet 模型的 FC2 層來代表影像特徵。

▓ 程式 3-7：利用 VGGNet16 預訓練模型的 FC2 層分析影像特徵

```java
package com.ai.deepsearch.features.deeplearning;

import org.datavec.image.loader.NativeImageLoader;
import org.deeplearning4j.nn.graph.ComputationGraph;
import org.deeplearning4j.util.ModelSerializer;
import org.nd4j.linalg.api.ndarray.INDArray;
import org.nd4j.linalg.dataset.api.preprocessor.DataNormalization;
import org.nd4j.linalg.dataset.api.preprocessor.VGG16ImagePreProcessor;

import java.io.File;
import java.io.IOException;
import java.util.Map;

/**
 *  利用 VGGNet16-ImageNet 資料集預訓練模型分析影像特徵
 */
public class VGG16Feature {
    // 載入預訓練的 VGG16 模型
    private ComputationGraph loadModel(String modelName) throws
```

```
IOException {
        File model = new File(modelName);
        if(model.exists()) {
            return ModelSerializer.restoreComputationGraph(model);
        } else {
            return null;
        }
    }
    // 利用 VGG16 模型 FC2 層來分析影像的特徵
    private INDArray getVgg16Feature(String imageName,ComputationGraph
model) throws IOException {
        File file=new File(imageName);
        NativeImageLoader loader=new NativeImageLoader(224,224,3);
        INDArray imageArray=loader.asMatrix(file);
        DataNormalization scaler=new VGG16ImagePreProcessor();
        scaler.transform(imageArray);
        Map<String,INDArray> map=model.feedForward(imageArray, false);
        INDArray feature=map.get("fc2");
        return feature;
    }

    public static void main(String[] args) {
        VGG16Feature modelFeature=new VGG16Feature();
        try {
            ComputationGraph vgg16=modelFeature.loadModel("resource/
vgg16_dl4j_inference.zip");
            if (vgg16!=null){
                INDArray feature=modelFeature.getVgg16Feature("resource/
image_name_ rgb8.jpg", vgg16);
                System.out.print("VGG16 模型分析特徵 :");
                System.out.println(feature.toString());
            } else {
```

```
              System.out.println(" 未找到 VGG16 模型檔案 !");
          }
      } catch (IOException e) {
          // TODO Auto-generated catch block
          e.printStackTrace();
      }
    }
}
```

3.3.5 使用移轉學習和微調技術進一步提升分析特徵 的精度

我們常常將以預訓練模型為基礎的影像檢索技術應用在某個垂直領域，有些領域的影像與預訓練資料集的影像類似，而有些領域的影像與預訓練資料集影像並不相似或類型迥異。對於前者，我們可以繼續使用 3.3.4 節中介紹的技術來分析影像特徵。而對於後者，如果我們不擁有垂直領域的巨量圖像資料集，就無法重新從頭訓練一個良好的模型來分析影像特徵，這時需要引用移轉學習和微調技術來改善所分析特徵的精度。當然無論對於前者還是後者，我們都可以使用此種方法來加強模型所分析特徵的精度。

從廣義上來說，移轉學習是一種能夠將在一個領域中學到的知識帶到新領域中的能力。移轉學習按照學習方式，又可以分為以樣本為基礎的移轉、以特徵為基礎的移轉、以模型為基礎的移轉，和以關係為基礎的移轉，這裡我們採用模型移轉的方式進行特徵分析能力的移轉學習。微調是指將已經訓練好的模型移轉到新領域的過程中，不必從頭訓練和最佳化參數，只需要對其參數做簡單調整即可進一步訓練新領域的模型。實際以什麼樣的方式實現模型的移轉與微調主要取決於兩點：新領域資料集的大小，以及新領域資料集與預訓練模型原始資料集的相似程度。按照這兩個因素的組合又會存在以下 4 種情形。

（1）當新領域資料集規模比較小並且與原始資料集比較相似時：由於新的資料集規模較小，微調旋積層的參數並不是一個明智的選擇，容易造成過擬合；由於新資料集與原始資料集比較相似，較高旋積層和全連接層的特徵也基本是相同的，所以只根據新資料集調整分類數量，進而訓練一個新的線性分類器無疑是最好的選擇。

（2）當新領域資料集規模比較大並且與原始資料集比較相似時：由於擁有更多的相關資料，所以我們可以微調所有層的參數，而不至於造成過擬合。

（3）當新領域資料集規模比較小但又與原始資料集存在很大不同時：因為資料集規模比較小，只訓練一個新的線性分類器無疑是最佳的。但新資料集與原始資料集又有很大的不同，只訓練分類器並不能表現新資料集的特徵，所以從某個更低的層開始訓練分類器更能表現資料的差異性。

（4）當新領域資料集規模很大並且與原始資料集差異較大時：由於新資料集的規模很大，所以可以從零開始訓練一個新的模型。然而即使是這樣，我們仍然能夠從模型移轉中獲益。通常使用預訓練的模型參數來初始化新模型的參數，並在此基礎上微調整個網路來訓練模型，進而加快模型的訓練。

在上一節中，提到了使用預訓練的 VGG16 模型 FC2 層來分析影像特徵的實例。那麼實際應該怎樣應用移轉學習和微調技術來提升分析特徵的精度呢？例如在某個領域內有影像搜索需求，我們要利用所擁有的該領域內的圖像資料集，對在 ImageNet 資料集上預訓練的 VGG16 模型進行移轉學習和參數微調，來進一步提升 FC2 層所分析特徵在該領域影像特徵上的轉換性。

假設該領域的圖像資料集 Example 存在 4 個類別的影像，程式 3-8 對該資料集進行了前置處理。程式 3-9 根據 Example 資料集具有 4 個分類影像的情況，將在 ImageNet 資料集 1000 個分類影像上預訓練的 VGG16 模型調整分類數量為 4，並將訓練後的模型輸出為 vgg16_dl4j_ finetune_last_layer.zip。程式 3-10 在調整分類數量後的模型 vgg16_dl4j_finetune_last_layer.zip 基礎上，繼續訓練微調所有全連接層的參數。

程式 3-8：某個領域的圖像資料集 Example

```java
package com.ai.deepsearch.deeplearning.datasets;

import org.datavec.api.io.filters.BalancedPathFilter;
import org.datavec.api.io.labels.ParentPathLabelGenerator;
import org.datavec.api.split.FileSplit;
import org.datavec.api.split.InputSplit;
import org.datavec.image.loader.BaseImageLoader;
import org.datavec.image.recordreader.ImageRecordReader;
import org.deeplearning4j.datasets.datavec.RecordReaderDataSetIterator;
import org.nd4j.linalg.dataset.api.DataSetPreProcessor;
import org.nd4j.linalg.dataset.api.iterator.DataSetIterator;
import org.nd4j.linalg.dataset.api.preprocessor.VGG16ImagePreProcessor;

import java.io.File;
import java.io.IOException;
import java.util.Random;

/**
 * 某個領域的圖像資料集
```

```java
*/
public class ExampleDataSetIterator {
    private static final String[] allowedExtensions = BaseImageLoader.
ALLOWED_FORMATS;
    private static final String DATASET = "resource/datasets/example";
    private static final Random rand = new Random(12);
    private static InputSplit trainSplit, testSplit;
    private static ParentPathLabelGenerator labelGenerator = new
ParentPathLabelGenerator();
    // 根據 Example 資料集中的影像類別數設定
    private static int numOfClasses = 4;

    private static DataSetIterator generateIterator(InputSplit
inputSplit, int batchSize) throws IOException {
        ImageRecordReader recordReader = new ImageRecordReader(224,
224, 3, labelGenerator);
        recordReader.initialize(inputSplit);
        DataSetIterator iterator = new RecordReaderDataSetIterator
(recordReader, batchSize, 1, numOfClasses);
        DataSetPreProcessor preProcessor = new VGG16ImagePreProcessor();
        iterator.setPreProcessor(preProcessor);
        return iterator;
    }

    // 將資料分為訓練集和驗證集
    public static void splitDataSet(int trainPercent) {
        File flowerDir = new File(DATASET);
        FileSplit files = new FileSplit(flowerDir, allowedExtensions,
rand);
        BalancedPathFilter balancedFilter = new BalancedPathFilter(rand,
allowedExtensions, labelGenerator);
        InputSplit[] filesSplited = files.sample(balancedFilter,
```

```
trainPercent, 100 - trainPercent);
        trainSplit = filesSplited[0];
        testSplit = filesSplited[1];
    }


    // 訓練集反覆運算器
    public static DataSetIterator trainIterator(int batchSize) throws
IOException {
        return generateIterator(trainSplit, batchSize);
    }


    // 測試集反覆運算器
    public static DataSetIterator testIterator(int batchSize) throws
IOException {
        return generateIterator(testSplit, batchSize);
    }
}
```

▌ 程式 3-9：訓練一個新的線性分類器

```
package com.ai.deepsearch.deeplearning.transfer;

import com.ai.deepsearch.deeplearning.datasets.ExampleDataSetIterator;
import org.deeplearning4j.eval.Evaluation;
import org.deeplearning4j.nn.api.OptimizationAlgorithm;
import org.deeplearning4j.nn.conf.Updater;
import org.deeplearning4j.nn.conf.distribution.NormalDistribution;
import org.deeplearning4j.nn.conf.layers.OutputLayer;
import org.deeplearning4j.nn.graph.ComputationGraph;
import org.deeplearning4j.nn.transferlearning.FineTuneConfiguration;
import org.deeplearning4j.nn.transferlearning.TransferLearning;
import org.deeplearning4j.nn.weights.WeightInit;
import org.deeplearning4j.util.ModelSerializer;
```

```java
import org.nd4j.linalg.activations.Activation;
import org.nd4j.linalg.dataset.api.iterator.DataSetIterator;
import org.nd4j.linalg.lossfunctions.LossFunctions;
import org.slf4j.Logger;
import org.slf4j.LoggerFactory;

import java.io.File;
import java.io.IOException;

/**
 * 訓練一個新的線性分類器
 */
public class FineTuneLastLayer {
    private static final Logger log = LoggerFactory
            .getLogger(FineTuneLastLayer.class);
    // 根據 Example 資料集中的影像類別數設定
    private int numOfClasses = 4;
    private int trainPercent = 80;
    private int batchSize = 15;


    // 載入預訓練的 VGG16 模型
    private ComputationGraph loadModel(String modelName) throws
IOException {
        File model = new File(modelName);
        if (model.exists()) {
            return ModelSerializer.restoreComputationGraph(model);
        } else {
            return null;
        }
    }


    // 微調模型最後一層
```

```java
    private boolean modifyLastLayer(ComputationGraph model) {
        // 微調參數設定
        FineTuneConfiguration fineTuneConf = new FineTuneConfiguration.
Builder()
                .learningRate(5e-5)
                .optimizationAlgo(
                    OptimizationAlgorithm.STOCHASTIC_GRADIENT_
DESCENT)
                .updater(Updater.NESTEROVS).seed(123456).build();

        // 模型移轉及微調設定
        ComputationGraph vgg16Transfer = new TransferLearning.
GraphBuilder(
                model)
                .fineTuneConfiguration(fineTuneConf)
                .setFeatureExtractor("fc2")
                .removeVertexKeepConnections("predictions")
                .addLayer(
                    "predictions",
                    new OutputLayer.Builder(
                        LossFunctions.LossFunction.
NEGATIVELOGLIKELIHOOD)
                        .nIn(4096)
                        .nOut(numOfClasses)
                        .weightInit(WeightInit.DISTRIBUTION)
                        .dist(new NormalDistribution(0,
                            0.2 * (2.0 / (4096 +
numOfClasses))))
                        .activation(Activation.SOFTMAX).
build(), "fc2")
                .build();
```

```java
// 準備資料
ExampleDataSetIterator.splitDataSet(trainPercent);
try {
    DataSetIterator trainIter = ExampleDataSetIterator
            .trainIterator(batchSize);
    DataSetIterator testIter = ExampleDataSetIterator
            .testIterator(batchSize);

    Evaluation eval;
    eval = vgg16Transfer.evaluate(testIter);
    log.info(" 評估 ");
    log.info(eval.stats() + "\n");
    testIter.reset();

    // 訓練
    int iter = 0;
    while (trainIter.hasNext()) {
        vgg16Transfer.fit(trainIter.next());
        if (iter % 10 == 0) {
            log.info(" 評估模型 , 第 " + iter + " 次反覆運算 ");
            eval = vgg16Transfer.evaluate(testIter);
            log.info(eval.stats());
        }
        iter++;
    }
    log.info(" 訓練完成 ");

    // 儲存訓練好的模型
    File file = new File("resource/vgg16_dl4j_finetune_last_
layer.zip");
    ModelSerializer.writeModel(vgg16Transfer, file, false);
    log.info(" 模型已儲存 ");
```

```
            return true;
        } catch (IOException e) {
            // TODO Auto-generated catch block
            e.printStackTrace();
            return false;
        }
    }
}
```

⌛ 程式 3-10：微調所有全連接層

```java
package com.ai.deepsearch.deeplearning.transfer;

import com.ai.deepsearch.deeplearning.datasets.ExampleDataSetIterator;
import org.deeplearning4j.eval.Evaluation;
import org.deeplearning4j.nn.conf.Updater;
import org.deeplearning4j.nn.graph.ComputationGraph;
import org.deeplearning4j.nn.transferlearning.FineTuneConfiguration;
import org.deeplearning4j.nn.transferlearning.TransferLearning;
import org.deeplearning4j.util.ModelSerializer;
import org.nd4j.linalg.dataset.api.iterator.DataSetIterator;
import org.slf4j.Logger;
import org.slf4j.LoggerFactory;

import java.io.File;
import java.io.IOException;

/**
 * 微調所有全連接層
 */
public class FineTuneAllFCLayer {
    private static final Logger log = LoggerFactory
            .getLogger(FineTuneAllFCLayer.class);
```

```
// 根據 Example 資料集中的影像類別數設定
private int numOfClasses = 4;
private int trainPercent = 80;
private int batchSize = 15;


// 載入預訓練的 VGG16 模型
private ComputationGraph loadModel(String modelName) throws
IOException {
    File model = new File(modelName);
    if (model.exists()) {
        return ModelSerializer.restoreComputationGraph(model);
    } else {
        return null;
    }
}


// 繼續微調所有全連接層
private boolean fineTuneAllFCLayer() throws IOException {
    // 載入修改最後一層已訓練的模型
    ComputationGraph model = loadModel("resources/vgg16_dl4j_
finetune_last_layer.zip");
    if (model != null) {
        // 微調參數設定
        FineTuneConfiguration fineTuneConf = new
FineTuneConfiguration.Builder()
                .learningRate(1e-5).updater(Updater.SGD).seed(123456)
                .build();
        // 模型移轉及微調設定
        ComputationGraph vgg16Transfer = new TransferLearning.
GraphBuilder(
                model).fineTuneConfiguration(fineTuneConf)
                .setFeatureExtractor("block4_pool").build();
```

```
// 準備資料
ExampleDataSetIterator.splitDataSet(trainPercent);
try {
    DataSetIterator trainIter = ExampleDataSetIterator
            .trainIterator(batchSize);
    DataSetIterator testIter = ExampleDataSetIterator
            .testIterator(batchSize);

    Evaluation eval;
    eval = vgg16Transfer.evaluate(testIter);
    log.info(" 評估 ");
    log.info(eval.stats() + "\n");
    testIter.reset();

    // 訓練
    int iter = 0;
    while (trainIter.hasNext()) {
        vgg16Transfer.fit(trainIter.next());
        if (iter % 10 == 0) {
            log.info(" 評估模型 , 第 " + iter + " 次反覆運算 ");
            eval = vgg16Transfer.evaluate(testIter);
            log.info(eval.stats());
        }
        iter++;
    }
    log.info(" 訓練完成 ");

    // 儲存訓練好的模型
    File file = new File(
            "resources/vgg16_dl4j_finetune_fc_layer_4.zip");
    ModelSerializer.writeModel(vgg16Transfer, file, false);
    log.info(" 模型已儲存 ");
```

```
        return true;
    } catch (IOException e) {
        // TODO Auto-generated catch block
        e.printStackTrace();
        return false;
    }
} else {
    System.out.println(" 未找到已修改最後一層的 VGG16 模型檔案 !");
    return false;
}
}
}
```

3.4 本章小結

目前，深度學習技術在影像特徵分析領域內所取得的成效已遠超傳統人工設計的方法。本章由什麼是深度學習談起，回顧了神經網路的發展史，介紹主要的神經網路模型和深度學習應用架構，並重點說明旋積神經網路各個理論要點，包含經典的旋積神經網路結構以及怎樣利用旋積神經網路分析影像特徵，最後介紹如何使用移轉學習和微調技術進一步提升分析特徵的精度。

04
Chapter

影像特徵索引
與檢索

4.1 影像特徵降維

第 2 章和第 3 章分別介紹以人工設計為基礎的影像特徵分析方法，和以深度學習為基礎的影像特徵分析方法。在這兩大類別圖像特徵中包含了許多高維度的特徵向量，例如 Gabor 小波特徵存在多個方向和尺度的特徵，組成了高維度的特徵向量；利用以深度學習為基礎的 AlexNet、VGGNet 等網路模型分析的特徵甚至高達 4096 維。高維度的影像特徵不僅帶來了較高的特徵向量儲存負擔，而且相當大地加強了特徵相似度比較的時間和空間複雜度。有沒有一種辦法可以在不影響結果正確性的基礎上降低這些負擔呢？在實際應用中，我們通常採用特徵選擇或特徵降維的方法來實現這一目的。特徵選擇是根據需要，從高維度的特徵中選擇其中的一部分作為新的特徵。特徵選擇使用嵌入式、過濾式、封裝式的方法，去除和最後結果不相關或相關度很低的特徵，保留高相關性的特徵。而特徵降維是利用高維度數據所固有的稀疏性，透過某種轉換將資料由高維空間對映到低維空間的方法。在影像特徵降維處理方法中，主成分分析（PCA）是最常用的演算法。另外，深度學習中的自動編碼器（Autoencoder）具有比主成分分析更好的降維效果，下面將詳細介紹這兩種方法。

4.1.1 主成分分析演算法降維

主成分分析演算法是由英國數學家 Karl Pearson 在 1901 年提出來的一種統計學方法，後來經過諸多科學家的進一步發展，逐步成為一種廣泛應用於資料統計和資料降維的演算法。使用主成分分析演算法實現資料降維的基本思維是透過某種線性投影，將資料由高維空間對映到低維空間，使對映後的資料間具有最大的離散度，並可以利用較少的資料維度來表達原有資料空間的特性。

主成分分析演算法通常採用將原始資料點分別向低維空間做投影的方法來實現對映，也就是 $Y=W^{\mathrm{T}}X$，其中 X 為原資料，Y 為投影後的資料，W^{T} 是投影矩陣。如何計算這個投影矩陣 W，使投影後的資料點盡可能分散，且投影後資料各維度間具有較低的相關性，就成為主成分分析演算法的核心問題。經過數學推導，我們需要找到這樣一個 W，使樣本點經過 W 投影後具有最大的方差，且各維度間兩兩協方差為 0。經過進一步演算，將 W 的計算簡化為對樣本的協方差矩陣進行特徵值分解的問題。

根據上面的思維，我們可以將主成分分析演算法降維過程歸納如下。

（1）每個樣本作為一個列向量，所有樣本組成一個樣本矩陣，其中每行代表一個維度。

（2）將樣本矩陣去中心化獲得矩陣 X，也就是每一維資料減去該維資料的平均值。

（3）計算樣本的協方差矩陣。

（4）計算協方差矩陣的特徵值和其對應的特徵向量。

（5）將特徵值按大小降冪排列，取其中的 k 個特徵值，並將它們對應的特徵向量組合成特徵矩陣 W。

（6）將資料 X 投影到低維空間，就獲得降維後的資料 Y，即 $Y=W^{\mathrm{T}}X$。

由於我們通常採用表格的形式來表示資料，為了與習慣相一致，我們將上述過程中的樣本矩陣和結果矩陣 Y 的組成方式改為：每個樣本作為一個行向量，每列為一維。因 $Y^{\mathrm{T}}=(W^{\mathrm{T}}X)^{\mathrm{T}}=X^{\mathrm{T}}W$，故投影的結果 = 輸入資料矩陣 × 特徵向量矩陣。

下面使用一個簡單的實例來具體地說明這個問題。假設現在有一組 3 維資料，如表 4-1 所示，每行是一個樣本，各列分別代表特徵 x、y、z。

表 4-1　一組 3 維資料

x	y	z
−1.0856306	0.99734545	0.2829785
−1.50629471	−0.57860025	1.65143654
−2.42667924	−0.42891263	1.26593626
−0.8667404	−0.67888615	−0.09470897
1.49138963	−0.638902	−0.44398196
−0.43435128	2.20593008	2.18678609
1.0040539	0.3861864	0.73736858
1.49073203	−0.93583387	1.17582904
−1.25388067	−0.6377515	0.9071052
−1.4286807	−0.14006872	−0.8617549

首先，需要將樣本資料中心化，經過計算 x、y、z 列的平均值分別為 −0.501608204、−0.0449493、0.6806994。中心化後的樣本矩陣如表 4-2 所示。

表 4-2 中心化後的資料

$x-\mu$	$y-\mu$	$z-\mu$
−0.584022	1.042295	−0.39772
−1.004687	−0.53365	0.970737
−1.925071	−0.38396	0.585237
−0.365132	−0.63394	−0.77541
1.9929978	−0.59395	−1.12468
0.0672569	2.250879	1.506087
1.5056621	0.431136	0.056669
1.9923402	−0.89088	0.49513
−0.752272	−0.5928	0.226406
−0.927072	−0.09512	−1.54245

接下來，計算樣本的協方差矩陣。在數學中通常使用協方差 $\text{Cov}(x,y) = \dfrac{\sum\limits_{i=1}^{n}(x_i - \bar{x})(sy_i - \bar{y})}{n-1}$ 表示二維資料的離散程度，其中 $\bar{x}$、$\bar{y}$ 分別表示各維資料平均值。對於多維資料，我們自然也就需要使用多個協方差來表示它們之間的關係，並使用協方差矩陣 $\boldsymbol{C}_{n \times n} = (c_{i,j}, c_{i,j} = \text{Cov}(Dim_i, Dim_j))$ 來表達。例如 3 維資料 x、y、z 的協方差矩陣表示為：

$$\boldsymbol{C} = \begin{bmatrix} \text{Cov}(x,x) & \text{Cov}(x,y) & \text{Cov}(x,z) \\ \text{Cov}(y,x) & \text{Cov}(y,y) & \text{Cov}(y,z) \\ \text{Cov}(z,x) & \text{Cov}(z,y) & \text{Cov}(z,z) \end{bmatrix}$$

表 4-1 中 3 維資料的協方差矩陣計算為：

$$\text{Cov} = \begin{bmatrix} 1.8697863 & -0.0806636 & -0.1550287 \\ -0.0806636 & 0.9644132 & 0.3320147 \\ -0.1550287 & 0.3320147 & 0.9173631 \end{bmatrix}$$

下一步，計算協方差矩陣的特徵值和特徵向量，獲得：

$$Eigenvalues = \begin{bmatrix} 0.6052503 & 1.2315945 & 1.9147178 \end{bmatrix}$$

$$Eigenvector = \begin{bmatrix} 0.047693 & -0.247817 & -0.967632 \\ -0.672240 & -0.724475 & 0.152410 \\ 0.738795 & -0.643212 & 0.201145 \end{bmatrix}$$

Eigenvalues 是協方差矩陣的特徵值，*Eigenvector* 是特徵值對應的特徵向量。特徵值 0.6052503、1.2315945、1.9147178 對應的特徵向量分別

是：$\begin{bmatrix} 0.047693 \\ -0.672240 \\ 0.738795 \end{bmatrix}$、$\begin{bmatrix} -0.247818 \\ -0.724475 \\ -0.643213 \end{bmatrix}$、$\begin{bmatrix} -0.967632 \\ 0.152410 \\ 0.201145 \end{bmatrix}$。特徵值按照由大到小的

順序排列，選取其中最大的 k 個特徵值，然後將其對應的特徵向量合併為一個特徵向量矩陣，k 就是要保留的維度。假如我們要將 3 維資料降為 2 維，結果如下：

$$\begin{bmatrix} -0.967632 & -0.247818 \\ 0.152410 & -0.724475 \\ 0.201145 & -0.643213 \end{bmatrix}$$

最後，將表 4-2 中樣本資料中心化後的矩陣與特徵向量矩陣相乘，獲得降維後的 2 維資料，如表 4-3 所示。

表 4-3　降維後的資料

a	b
−0.6439752	−0.3545665
−1.0860922	0.0112055
−1.9219582	0.3788066
−0.1007258	1.0485100
2.2452371	0.6598132
−0.5809183	−2.6161072

a	b
1.3798189	−0.7219269
1.9640391	−0.1667869
−0.6831145	0.4702697
−0.5723108	1.2907824

計算矩陣特徵值和特徵向量的方法（EVD）由於計算量較大，並不適用於大規模資料的降維處理。另外還有一種方法，它利用了矩陣的奇異值分解（SVD），任何分解 $A=U\sum V^T$ 稱為矩陣 A 的奇異值分解，其中 U 和 V 是正交矩陣，$\sum$ 是一個形如 $\begin{bmatrix} D & 0 \\ 0 & 0 \end{bmatrix}$ 的矩陣 $\begin{bmatrix} \begin{bmatrix} \sigma_1 & \cdots & 0 \\ \vdots & & \vdots \\ 0 & \cdots & \sigma_r \end{bmatrix} & 0 \\ 0 & 0 \end{bmatrix}$。$\sigma_1$，$\sigma_2$，$\cdots$，$\sigma_r$ 被稱為奇異值，並且按照數值大小降冪排列。我們可以選取前 k 個奇異值，忽略剩下的 $r-k$ 個奇異值，重新組成 D，這樣就保留了原有資料的大部分特性，縮減後的$\sum$所對應的 V 就是投影矩陣。

▓ 程式 4-1：PCA 降維的實現

```java
package com.ai.deepsearch.index;

import Jama.EigenvalueDecomposition;
import Jama.Matrix;
import Jama.SingularValueDecomposition;

/**
 * PCA 降維
 */
public class PCA {
    // 測試資料
    public double[][] testData = {{-1.0856306, 0.99734545, 0.2829785},
            {-1.50629471, -0.57860025, 1.65143654},
```

```
        {-2.42667924, -0.42891263, 1.26593626},
        {-0.8667404, -0.67888615, -0.09470897},
        {1.49138963, -0.638902, -0.44398196},
        {-0.43435128, 2.20593008, 2.18678609},
        {1.0040539, 0.3861864, 0.73736858},
        {1.49073203, -0.93583387, 1.17582904},
        {-1.25388067, -0.6377515, 0.9071052},
        {-1.4286807, -0.14006872, -0.8617549}};


// 實現 PCA 的兩種方法
public enum PCAMethod {
    SVD, EVD
}


// 矩陣資料中心化
private double[][] getCenteredData(double[][] data) {
    // 行數
    int m = data.length;
    // 列數
    int n = data[0].length;
    // 維度數據平均值
    double[] avg = new double[n];
    // 資料中心化後的矩陣
    double[][] centeredData = new double[m][n];
    // 計算維度平均值
    for (int i = 0; i < n; i++) {
        double sum = 0;
        for (int j = 0; j < m; j++) {
            sum += data[j][i];
        }
        avg[i] = sum / m;
    }
    // 減去平均值
```

```java
        for (int i = 0; i < n; i++) {
            for (int j = 0; j < m; j++) {
                centeredData[j][i] = data[j][i] - avg[i];
            }
        }
        return centeredData;
    }

    // 計算協方差矩陣
    private Matrix getCovMatrix(double[][] centeredData) {
        // 行數
        int m = centeredData.length;
        // 列數
        int n = centeredData[0].length;
        // 協方差矩陣
        double[][] covData = new double[n][n];
        for (int i = 0; i < n; i++) {
            for (int j = 0; j < n; j++) {
                double sum = 0;
                for (int k = 0; k < m; k++) {
                    sum += centeredData[k][i] * centeredData[k][j];
                }
                covData[i][j] = sum / (m - 1);
            }
        }
        return new Matrix(covData);
    }

    // 線性代數方法求協方差矩陣   Cov=A'A/(n-1)
    private Matrix getCovMatrixByLA(double[][] centeredData) {
        Matrix centeredMatrix = new Matrix(centeredData);
        Matrix centeredMatrixT = centeredMatrix.transpose();
        Matrix covMatrix = centeredMatrixT.times(centeredMatrix).times
```

```
(1.0 / (centeredData.length - 1));
        return covMatrix;
    }


    /*
     * 使用奇異值計算
     * pComponents 表示保留的主成分數。當 pComponents 為零時，採用百分比設定值
的方式。
     * threshold 表示保留主成分的比例。當 threshold 為零時，採用數量設定值的方式。
     */
    private Matrix svdPCA(double[][] data, int pComponents, double
threshold) {
        Matrix dataMatrix = new Matrix(getCenteredData(data));
        SingularValueDecomposition svd = dataMatrix.svd();
        // 計算奇異值 sigma
        double[] sigma = svd.getSingularValues();
        // lambda=sigma*sigma
        double[] lambda = new double[sigma.length];
        double sum = 0;
        // 計算對角陣
        Matrix diagonalMatrix = svd.getS();
        // 對角陣行數和列數
        int rows = diagonalMatrix.getRowDimension();
        int cols = diagonalMatrix.getColumnDimension();
        Matrix subMatrix = null;
        if (pComponents == 0) {
            for (int i = 0; i < cols; i++) {
                lambda[i] = sigma[i] * sigma[i];
                sum += lambda[i];
            }
            int i = 0;
            double rate = lambda[0] / sum;
            while (rate <= threshold) {
```

```
                    rate += lambda[++i] / sum;
            }
        submatrix = svd.getV().getMatrix(0, rows - 1, 0, i);
    }
    if (threshold == 0) {
        submatrix = svd.getV().getMatrix(0, rows - 1, 0,
pComponents - 1);
    }


    return submatrix;
}

/*
* 使用特徵向量計算
* pComponents 表示保留的主成分數。當 pComponents 為零時，採用百分比設定值的
方式。
* threshold 表示保留主成分的比例。當 threshold 為零時，採用數量設定值的方式。
*/
private Matrix evdPCA(double[][] data, int pComponents, double
threshold) {
    double[][] centeredData = getCenteredData(data);
    // 協方差矩陣
    Matrix covMatrix = getCovMatrixByLA(centeredData);
    EigenvalueDecomposition evd = covMatrix.eig();
    // 特徵值矩陣
    Matrix eigenvalueMatrix = evd.getD();
    // 特徵值之和
    double sum = eigenvalueMatrix.trace();
    double rate = 0;
    int cols = eigenvalueMatrix.getColumnDimension();
    int rows = eigenvalueMatrix.getRowDimension();
    Matrix eigenVectorMatrix = evd.getV();
    Matrix submatrix = null;
```

```java
        if (pComponents == 0) {
            int i = cols;
            // 取設定值之內的特徵值
            while (rate <= threshold) {
                i--;
                rate += eigenvalueMatrix.get(i, i) / sum;
            }
            int[] indices = new int[cols - i];
            for (int j = 0; j < cols - i; j++) {
                indices[j] = cols - 1 - j;
            }
            subMatrix = eigenVectorMatrix.getMatrix(0, rows - 1, indices);
        }
        if (threshold == 0) {
            subMatrix = eigenVectorMatrix.getMatrix(0, rows - 1, cols -
1, cols -  pComponents);
        }

        return subMatrix;
    }


    public Matrix reduceDims(PCAMethod method, double[][] data, int
pComponents, double threshold) {
        System.out.println(" 原始資料 ");
        new Matrix(data).print(9, 7);
        // 行數
        int rows = data.length;
        // 列數
        int cols = data[0].length;
        // 樣本數小於維度數，不能處理
        if (rows < cols) {
            return null;
        }
```

```java
        // 中心化資料
        Matrix dataMatrix = new Matrix(getCenteredData(testData));
        Matrix pcaMatrix;
        switch (method) {
            case SVD:
                pcaMatrix = svdPCA(data, pComponents, threshold);
            case EVD:
                pcaMatrix = evdPCA(data, pComponents, threshold);
            default:
                pcaMatrix = svdPCA(data, pComponents, threshold);
        }
        return dataMatrix.times(pcaMatrix);
    }


    public static void main(String args[]) {
        PCA pca = new PCA();
        // 特徵值方法，百分比設定值
        Matrix evdReduceMatrix0 = pca.reduceDims(PCAMethod.EVD, pca.
testData, 0, 0.80);
        System.out.println("降維後的資料");
        evdReduceMatrix0.print(9, 7);
        // 特徵值方法，數值設定值
        Matrix evdReduceMatrix1 = pca.reduceDims(PCAMethod.EVD, pca.
testData, 2, 0);
        System.out.println("降維後的資料");
        evdReduceMatrix1.print(9, 7);
        // 奇異值方法，百分比設定值
        Matrix svdReduceMatrix0 = pca.reduceDims(PCAMethod.SVD, pca.
testData, 0, 0.80);
        System.out.println("降維後的資料");
        svdReduceMatrix0.print(9, 7);
        // 奇異值方法，數值設定值
        Matrix svdReduceMatrix1 = pca.reduceDims(PCAMethod.SVD, pca.
```

```
testData, 2, 0);
        System.out.println(" 降維後的資料 ");
        svdReduceMatrix1.print(9, 7);
    }
}
```

4.1.2 深度自動編碼器降維

自動編碼器（AutoEncoder）是一種可以對輸入樣本進行壓縮並近似還原的神經網路。最簡單的自動編碼器可以表示為圖 4-1 的樣子，左邊是輸入層，右邊是輸出層，中間是一個全連接的隱藏層。可以看到，隱藏層的神經元個數明顯比輸入層要少，這樣做實際上是對輸入樣本進行了壓縮。而輸出層的神經元個數與輸入層保持一致，也就是將壓縮後的樣本進行還原。由輸入層到隱藏層的過程，稱之為編碼（Encoder）；由隱藏層到輸出層的過程，稱之為解碼（Decoder）。開發過程能夠明顯地壓縮樣本資料的大小，進一步實現資料降維的作用。

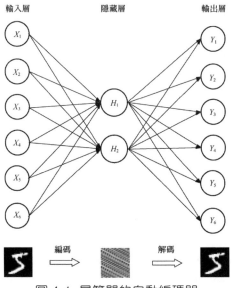

圖 4-1 最簡單的自動編碼器

自動編碼器在經過某種資料的訓練後，對這種資料會有很好的資料降維效果。對於其他未經訓練的類別資料，自動編碼器常常表現地不盡如人意。隨著深度的增加，神經網路會表現出更好的效能。深度自動編碼器（Deep AutoEncoder）是由 Geoff Hinton 提出的一種用於資料壓縮、降維的深度神經網路[1]。如圖 4-2 所示，它由兩個對稱的深信度網路（DBN）組成，一個負責編碼，另一個負責解碼。每個深信度網路又由 4 ～ 5 個淺層網路堆疊而成，每個淺層網路都是一個受限玻爾茨曼機（RBM），它是組成深信度網路的基本單元。受限玻爾茨曼機是一種只有兩層的淺層神經網路，第一層是輸入層，第二層是隱藏層，兩層之間全連接。它之所以被稱為受限的，是因為同一層的神經元間並不相連。

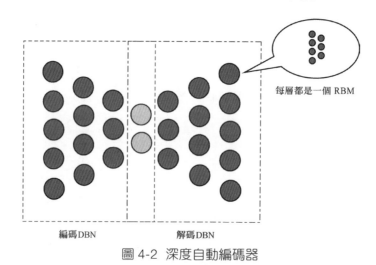

每層都是一個 RBM

編碼DBN　　　解碼DBN

圖 4-2　深度自動編碼器

程式 4-2 是深度自動編碼器的實作方式，編碼部分將輸入由 $width \times height$ 維逐步壓縮為 $1000 \rightarrow 500 \rightarrow 250 \rightarrow 100 \rightarrow 30$ 維大小，

1　Hinton G E,Salakhutdinov R R. Reducing the dimensionality of data with neural networks[J]. Science, 2006, 313(5786): 504-507.

解碼部分又將 30 維的壓縮資料逐步還原為 100 → 250 → 500 → 1000 → width×height 大小。當需要利用深度自動編碼器對特徵資料進行降維時，首先要將該資料類型的特徵資料集輸入 DeepAutoEncoderModel，並進行訓練；待模型訓練到理想的降維效果時，就可以使用 List<INDArray> model.feedForwardToLayer(int layerNum，INDArray input) 方法，來取出輸入特徵資料 input 的降維結果。

⧗ 程式 4-2：深度自動編碼器

```java
package com.ai.deepsearch.deeplearning.models;

import org.deeplearning4j.nn.api.OptimizationAlgorithm;
import org.deeplearning4j.nn.conf.MultiLayerConfiguration;
import org.deeplearning4j.nn.conf.NeuralNetConfiguration;
import org.deeplearning4j.nn.conf.layers.OutputLayer;
import org.deeplearning4j.nn.conf.layers.RBM;
import org.deeplearning4j.nn.multilayer.MultiLayerNetwork;
import org.nd4j.linalg.activations.Activation;
import org.nd4j.linalg.lossfunctions.LossFunctions;

/**
 * 深度自動編碼器模型
 */
public class DeepAutoEncoderModel {
    private int width;
    private int height;
    private int seed = 123;
    private int iterations = 1;

    public DeepAutoEncoderModel(int width, int height, int seed, int iterations) {
        this.width = width;
```

```
            this.height = height;
            this.seed = seed;
            this.iterations = iterations;
    }


    public MultiLayerNetwork initModel() {
        MultiLayerConfiguration conf = new NeuralNetConfiguration.
Builder()
                .seed(seed)
                .iterations(iterations)
                .optimizationAlgo(OptimizationAlgorithm.LINE_GRADIENT_
DESCENT)
                .list()
                // 編碼部分  width*height->1000->500->250->100->30
                .layer(0, new RBM.Builder().nIn(width * height).
nOut(1000).lossFunction(LossFunctions.LossFunction.KL_DIVERGENCE).
build())
                .layer(1, new RBM.Builder().nIn(1000).nOut(500).
lossFunction(LossFunctions.LossFunction.KL_DIVERGENCE).build())
                .layer(2, new RBM.Builder().nIn(500).nOut(250).
lossFunction(LossFunctions.LossFunction.KL_DIVERGENCE).build())
                .layer(3, new RBM.Builder().nIn(250).nOut(100).
lossFunction(LossFunctions.LossFunction.KL_DIVERGENCE).build())
                .layer(4, new RBM.Builder().nIn(100).nOut(30).
lossFunction(LossFunctions.LossFunction.KL_DIVERGENCE).build())
                // 解碼部分  30->100->250->500->1000->width*height
                .layer(5, new RBM.Builder().nIn(30).nOut(100).
lossFunction(LossFunctions.LossFunction.KL_DIVERGENCE).build())
                .layer(6, new RBM.Builder().nIn(100).nOut(250).
lossFunction(LossFunctions.LossFunction.KL_DIVERGENCE).build())
                .layer(7, new RBM.Builder().nIn(250).nOut(500).
lossFunction(LossFunctions.LossFunction.KL_DIVERGENCE).build())
                .layer(8, new RBM.Builder().nIn(500).nOut(1000).
```

```
lossFunction(LossFunctions.LossFunction.KL_DIVERGENCE).build())
            .layer(9, new OutputLayer.Builder(LossFunctions.
LossFunction.MSE).activation(Activation.SIGMOID).nIn(1000).nOut(width *
height).build())
            .pretrain(true).backprop(true)
            .build();

    // 根據設定建置網路模型
    MultiLayerNetwork deepAutoEncoderModel = new
MultiLayerNetwork(conf);
    deepAutoEncoderModel.init();

    return deepAutoEncoderModel;
    }
}
```

4.2 影像特徵標準化

為了使不同量綱的特徵能夠相互比較，我們通常需要將這些特徵資料進行標準化。特徵資料的標準化將資料等比例縮放到一定的數值區間，消除各種量綱的影響，便於影像特徵的比較。常用的標準化方法有離差標準化和標準差標準化等方法。

4.2.1 離差標準化

離差標準化又被稱為 Min-Max 標準化，它對資料進行線性轉換，使轉換後的結果處於 [0,1] 區間。轉換公式如下，其中 *max* 為資料的最大值，*min* 為資料的最小值。

$$x' = \frac{x - min}{max - min} \tag{4-1}$$

⧗ 程式 4-3：離差標準化函數

```java
private void minmaxNormalize(double[] feature) {
    double min = Double.MAX_VALUE, max = Double.MIN_VALUE;
    for (int i = 0; i < feature.length; i++) {
        min = Math.min(feature[i], min);
        max = Math.max(feature[i], max);
    }
    for (int i = 0; i < feature.length; i++) {
        feature[i] = (feature[i] - min) / (max - min);
    }
}
```

4.2.2 標準差標準化

標準差標準化又被稱為 Z-Score 標準化，經它處理的資料符合標準正態分佈。轉換公式如下，其中 μ 為平均值，σ 為標準差。

$$x' = \frac{x - \mu}{\sigma} \tag{4-2}$$

⧗ 程式 4-4：標準差標準化函數

```java
private void zscoreNormalize(double[] feature) {
    double sumOfSquares = 0;
    double sum = 0;
    for (double each : feature) {
        sumOfSquares += (each * each);
        sum += each;
    }
    double mean = sum / feature.length;
```

```
if (sumOfSquares > 0) {
    sumOfSquares = Math.sqrt(sumOfSquares);
    for (int i = 0; i < feature.length; i++) {
        feature[i] = (feature[i] - mean) / sumOfSquares;
    }
}
```

4.3 影像特徵相似度的度量

目前我們已經能夠分析影像的特徵,並對高維的影像特徵進行降維和標準化。面對相似的兩個影像,要怎樣才能在特徵上度量它們的相似度呢?我們通常將影像特徵表示為向量的形式,因此判斷兩幅影像特徵的相似度就轉化為向量空間中兩點間距離的問題,接著通常會採用歐氏距離、曼哈頓距離、漢明距離、餘弦相似度、傑卡德相似度等指標來衡量向量間的相似度。

4.3.1 歐氏距離

歐氏距離全稱為阿基米德距離,通常用於計算歐氏空間中兩點間的距離。假設 $a(a_1, a_2, \cdots, a_n)$ 和 $b(b_1, b_2, \cdots, b_n)$ 是 n 維空間中的兩點,它們之間的歐氏距離可以表示為:

$$Distance(\boldsymbol{a}, \boldsymbol{b}) = \sqrt{\sum_{i=1}^{n}(a_i - b_i)^2} \tag{4-3}$$

很明顯,$n = 2$ 時的歐氏距離也就是曾經在平面幾何中所學的 $A(x_1, y_1)$、$B(x_2, y_2)$ 兩點間的距離 $|AB| = \sqrt{(x_1 - x_2)^2 + (y_1 - y_2)^2}$。

⧗ 程式 4-5：歐氏距離

```java
private double euclideanDistance(double[] a, double[] b) {
    assert (a.length == b.length);
    double sum = 0;
    for (int i = 0; i < a.length; i++) {
        sum += (a[i] - b[i]) * (a[i] - b[i]);
    }
    return Math.sqrt(sum);
}
```

4.3.2 曼哈頓距離

曼哈頓距離的名字，來自類似美國曼哈頓的區塊城市街區間最短行車路徑的計算方式。曼哈頓距離使用標準座標系上的絕對軸距總和表示，它比歐氏距離的計算量少，效能更好。曼哈頓距離的計算公式為：

$$Distance(\boldsymbol{a},\boldsymbol{b})=\sum_{i=1}^{n}\left|a_i - b_i\right| \tag{4-4}$$

⧗ 程式 4-6：曼哈頓距離

```java
private double manhattanDistance(double[] a, double[] b) {
    assert (a.length == b.length);
    double sum = 0;
    for (int i = 0; i < a.length; i++) {
        sum += Math.abs(a[i] - b[i]);
    }
    return sum;
}
```

4.3.3 漢明距離

漢明距離因它的提出者 Richard Hamming 而得名，主要應用於資訊理論、編碼論和密碼學中。漢明距離表示兩個字串（等長）在對應位置不同字元的個數。例如字串 a 為 111101，b 為 101111，則它們的漢明距離為 2。

▌ 程式 4-7：漢明距離

```java
private double hammingDistance(double[] a, double[] b) {
    assert (a.length == b.length);
    int distance = 0;
    for (int i = 0; i < a.length; i++) {
        if (a[i] != b[i]) distance++;
    }
    return distance;
}
```

4.3.4 餘弦相似度

餘弦相似度又稱為餘弦距離，使用 n 維向量空間中的兩個向量間的夾角的餘弦值來表示向量間的相似性。根據二維空間中向量 a、b 的點積公式 $a \cdot b = \|a\|\|b\|\cos\theta$ ，可推知 $\cos\theta = \dfrac{a \cdot b}{\|a\|\|b\|}$ 。假設向量 a 和 b 的座標分別為

(x_1, y_1)、(x_2, y_2)，那麼 $\cos\theta = \dfrac{x_1 x_2 + y_1 y_2}{\sqrt{x_1^2 + y_1^2} \times \sqrt{x_2^2 + y_2^2}}$ 。將之推廣到 n 維空間，向

量 $A(a_1, a_2, \cdots, a_n)$、$B(b_1, b_2, \cdots, b_n)$ 間夾角的餘弦值為 $\cos\theta = \dfrac{\sum\limits_{i=1}^{n}(A_i \times B_i)}{\sqrt{\sum\limits_{i=1}^{n} A_i^2} \times \sqrt{\sum\limits_{i=1}^{n} B_i^2}}$ 。

由於向量間夾角越小，其餘弦值越大，向量間的相似度也就越大，故向量間夾角的餘弦值和向量相似度成正比。

⟁ 程式 4-8：餘弦相似度

```java
private double cosineSimilarity(double[] a, double[] b) {
    assert (a.length == b.length);
    double distance = 0;
    double sumOfSquare1 = 0;
    double sumOfSquare2 = 0;
    for (int i = 0; i < a.length; i++) {
        distance += a[i] * b[i];
        sumOfSquare1 += a[i] * a[i];
        sumOfSquare2 += b[i] * b[i];
    }
    return distance / (Math.sqrt(sumOfSquare1) * Math.sqrt(sumOfSquare2));
}
```

4.3.5 傑卡德相似度

傑卡德相似度由 Paul Jaccard 提出，它代表了兩個集合的相似程度。集合 A 與集合 B 的傑卡德相似度可以定義為 A 與 B 的交集和其聯集大小之間的比值，即：

$$JAC(A, B) = \frac{|A \cap B|}{|A \cup B|} \tag{4-5}$$

如圖 4-3 所示，A 與 B 的交集中有 2 個元素，聯集中有 9 個元素，那麼它們的傑卡德相似度為 $JAC(A, B)$=2/9。可以想像一下：當 A 和 B 相互遠離而不相交時，兩個集合中相同資料的數量為 0，也就是 $JAC(A, B)$=0；當 A 和 B 相互接近而完全重合時，兩個集合中相同資料的數量是一樣的，也就是 $JAC(A, B)$=1；所以 $JAC(A, B) \in [0, 1]$。

圖 4-3 集合 A 與 B 的傑卡德相似度

⚡ 程式 4-9：傑卡德相似度

```java
private double jaccardSimilarity(double[] a, double[] b) {
    assert (a.length == b.length);
    // A和B的交集元素個數
    int intersection = 0;
    Set<Double> A = new HashSet<>();
    Set<Double> B = new HashSet<>();
    // A和B的聯集
    Set<Double> union = new HashSet<>();
    for (double each : a) {
        A.add(each);
    }
    for (double each : b) {
        if (A.contains(each)) {
            intersection++;
        }
        B.add(each);
    }
    union.addAll(A);
    union.addAll(B);

    return (double) intersection / union.size();
}
```

4.4 影像特徵索引與檢索

在掌握了影像特徵相似度的度量方法後，我們距離最後實現一個影像搜尋引擎的目標更近了一步。影像搜尋引擎將查詢影像與影像函數庫中的影像進行特徵相似度比較後，最後傳回許多幅相似影像的過程其實是一個 KNN（K 最近鄰）尋找問題。

4.4.1 從最近鄰（NN）到 K 最近鄰（KNN）

假設現在有一個已經分類的影像函數庫，如何來判斷一幅未知類別的影像屬於影像函數庫中的哪一種呢？一個最容易想到、最直接的辦法就是將這幅影像與影像函數庫中影像進行特徵比較，計算相似度，相似度最大的那幅影像的類別就是該幅影像的類別。這就是最近鄰（Nearest Neighbor）的思維，簡單而純粹，但是它也存在一定的問題。下面來觀察圖 4-4 中的內容，這個未知圖形究竟屬於哪一種，是三角形還是正方形呢？按照最近鄰的思維，它應該屬於正方形，但它不遠處還有更多的三角形，這怎麼解釋呢？其實單純依靠最近鄰去判斷類別會存在很大程度的誤判，在實際應用中常常會取得差強人意的結果，因此我們還需要結合周圍的情況來分析。K 最近鄰（K Nearest Neighbor）結合未知樣本周圍 K 個最近鄰的情況去判斷它的類別，這樣進一步加強了類別判斷的準確性。

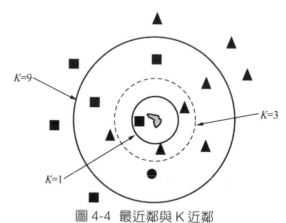

圖 4-4 最近鄰與 K 近鄰

KNN 同樣採用將未知數據與資料集中已標記的資料進行比較的策略，找到資料集中前 K 個最為相似的資料，並統計 K 個資料所屬的類別，出現次數最多的類別就是未知數據的類別。在圖 4-4 中，當 $K=1$ 時，很明顯，未知數據屬於正方形，這時實際上採用的是最近鄰（NN）演算法；當 $K=3$ 時，三角形的個數是 2，正方形的個數是 1，未知數據屬於三角形；當 $K=9$ 時，三角形的個數是 5，未知數據是三角形。

同樣，由於人們對相似影像的了解並不一樣，為了保持良好的使用者體驗，通常影像搜尋引擎不只傳回一幅「最像」的影像，而是傳回許多幅相似影像，並按一定的相似度排序。

4.4.2 索引建置與檢索

線性比較是 KNN 最簡單的實現方式，當資料規模較小時，它簡單而有效。但當面對巨量資料時，這種實現方式卻因效率太低而無法使用。為了解決這一問題，研究人員提出了許多可行的方法。在這些方法中，有的利用特殊的資料結構，有的利用資料本身所呈現的簇狀集聚特性，有的利用特殊的對映演算法。其實所有這些方法背後的策略都是首先對資料空間進行劃分，分成許多相似資料聚集的小空間，檢索時能夠使用某種方法直接定位到該空間，然後在小空間中做相似度的比較，減少比較次數，縮短檢索時間，相當大地提高檢索效率。

在實際專案實作中，能夠實現這一策略的成熟方案又分為以樹結構為基礎的方法、以向量量化為基礎的方法，和局部敏感雜湊方法這 3 大類，下面對它們逐一介紹。

◼ 以樹結構為基礎的索引建置與檢索

以樹結構方法為基礎的思維是將每個資料視為樹的節點，將這些節點建置為一棵二叉尋找樹，利用二叉尋找樹折半尋找的優勢來減少查詢時間。

K-d tree 是一種典型的以樹結構為基礎的方法，由史丹佛大學的 Jon Louis Bentley 在 1975 年提出[2]。K-d tree 中的 d 是 dimension 的縮寫，K 代表維度數，也就是說 K-d tree 是一種將許多個資料點劃分到 K 維空間的樹狀資料結構。為了讓讀者更直觀地來了解，下面看一個維基百科上的實例。如圖 4-5 所示，二維空間中有（2，3）、（5，4）、（9，6）、（4，7）、（8，1）、（7，2）這 6 個點，我們如何對它們進行空間劃分，進而建置一棵 K-d tree 呢？

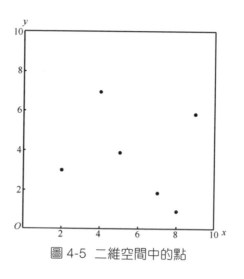

圖 4-5　二維空間中的點

首先要確定在哪個維度上對空間進行劃分，為了優先選擇區分度大的維度，需要比較各維度的方差。經計算，x 維度的方差明顯大於 y 維度，那麼將資料按照 x 維的大小昇冪排列並取中間值 7，使用 $x=7$ 對二維空間進行切分，並將點（7，2）作為根節點，切分點（7，2）前面的點作為左子樹，後面的點作為右子樹，進一步形成一個平衡的 K-d tree。接

2　Bentley JL.Multidimensional binary search trees used for associative searching[M]. ACM,1975.

下來，繼續按照相同的方法對左右子空間進行劃分。經計算，y 維度的方差大於 x 維度的方差，我們將資料按照 y 維昇冪排列取中間值 4，使用 $y=4$ 對左子空間繼續切分，並將點（5，4）作為左子樹的根節點。同理，使用 $y=6$ 對右子空間切分，並將點（9，6）作為右子樹的根節點。K-d tree 的劃分維度按照樹的層次循環選擇，剩餘的空間會按照 x 維繼續劃分，最後形成圖 4-6 所示的結果。

由上面的實例，可以歸納出建置 K-d tree 索引樹的演算法。

第一步：在一個 K 維的資料集合中選擇方差最大的維度 K_i，在該維度上對資料進行降冪排列並取它的中間值 mid，在 $K_i=mid$ 處使用垂直於 K_i 的超平面，將資料集對應的多維空間劃分為左右兩個子空間（K_i 維資料小於 mid 的劃為左子空間，大於 mid 的劃為右子空間），同時建立根節點 node 用於儲存 mid 對應的資料值。

第二步：對劃分出的兩個子空間重複第一步的過程，直到所有子空間不能再劃分為止，並將該子空間中的剩餘資料儲存到建立的葉子節點。

建置 K-d tree 索引樹的過程就是使用樹結構將資料進行索引的過程。那麼下一步，該如何在剛才建置的 K-d tree 上進行資料檢索呢？例如要查詢點（2.3，3.2）在該樹中的最近鄰點（如圖 4-7 所示）：點（2.3，3.2）在 x 維上明顯小於點（7，2），進入左子樹，按逐層進行空間劃分的維度進行二叉尋找，最後到達點（2，3），將該點優先考慮為最近鄰點。計算點（2.3，3.2）與點（2，3）的距離為 0.3606，以點（2.3，3.2）為圓心，0.3606 為半徑畫圓，該圓明顯不與 $y=4$ 的超平面相交，故不用考慮點（5，4）的右子樹，更不用考慮點（7，2）的右子樹，這樣點（2，3）就是最後確定的最近鄰點。以上實例只比較了 2 個點就找到最近鄰點，充分説明樹狀結構折半尋找的優勢。

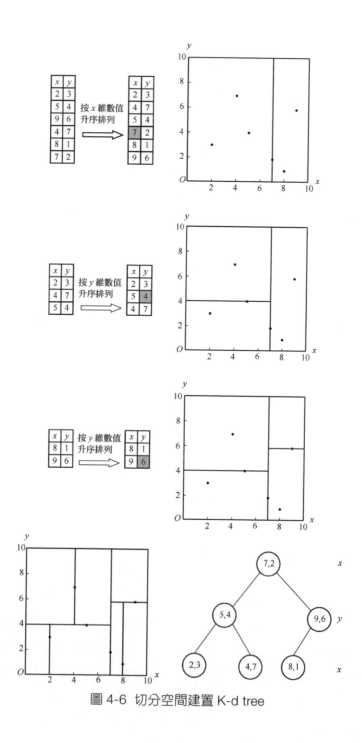

圖 4-6 切分空間建置 K-d tree

K-d tree 最近鄰查詢演算法可以歸納如下。首先，從根節點開始遞迴地向下比較查詢資料 Q 和目前節點 C 在空間切分維度 d 的大小。若 $Q_d<C_d$，則繼續比較左子樹，否則比較右子樹。直到節點 C 為葉節點為止，並將 C 作為最近鄰並記為 N。然後沿比較路徑遞迴地傳回，並對回復的目前節點 C 進行以下操作：如果 Q 的資料與目前節點 C 的資料間的距離 D_{QC} 小於 Q 與 N 資料的距離 D_{QN}，那麼將目前節點 C 記為最近鄰 N；如果目前節點 C 的兄弟節點 B 對應的空間與以 Q 為球心，以 Q 與目前最近鄰 N 的距離 D_{QN} 為半徑的超球體相交，那麼在這一空間有可能會存在距離 Q 更近的點，這時將目前節點移動到該節點 B 繼續遞迴地進行最近鄰查詢。如果不相交，則繼續向上回復。若目前節點退到根節點時，查詢過程結束，最後的最近鄰點 N 就是最近鄰點。

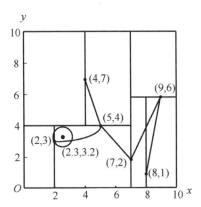

圖 4-7　K-d tree 的最近鄰查詢範例

雖然以 K-d tree 為代表的樹狀結構能夠加強資料查詢的效率，但許多對它效能基準測試的結果表明當資料的維度大於 10 時，K-d tree 的查詢效率會急遽下降，當資料的維度進一步增加時，甚至它的效率還不如線性比較。為此不少研究者提出了許多不同的解決方法，有的方法進一步最佳化了樹的局部結構減少比較的次數，例如球樹；有的方法在最佳化演算法的基礎上採用設定值策略進一步限制比較次數，以犧牲查詢精度為代價獲得近似最近鄰（ANN），例如 1997 年 David Lowe 提出的 BBF（Best Bin First）查詢演算法[3]。這些改進的結構和演算法雖然在某種程度上緩解了「維度災難」所帶來的挑戰，實現進一步加強高效查詢的資料維度，但是由於樹狀結構本身所固有的特性，在面對成百上千的資料維度時，這些方法依然收效甚微。

2 以向量量化方法為基礎的索引建置與檢索

在第 2 章介紹數字影像時曾說明了量化的方法，例如將影像量化為 8 位元色彩就是用 256 個色彩值來表示連續的像素值，也就是將連續的像素值平均分為 256 個區域，每個區域用一個色彩值表示，然後根據像素值與每個區域的距離使用對應的色彩值來表示。向量量化（Vector Quantization）就是將向量空間中的點用一個有限向量子集 Y 來表示，並保障失真最小的過程。這裡的 Y 稱為碼書，碼書中的元素稱為代碼，向量量化又可以視為將向量用代碼重新編碼的過程。

向量量化的研究源於 20 世紀 80 年代，它主要應用於數位通訊系統的信源編碼。由於它具有良好的壓縮編碼特性，人們開始將其用於影像檢索領域。向量量化技術應用於影像檢索領域最為經典的實例是視覺詞袋模型（Bags of Visual Word，BOVW）。如圖 4-8 所示，視覺詞袋模型首先分析影像函數庫中所有影像的特徵，然後利用某種分群演算法（一般使用 K-means）將全部影像特徵分群，並使用分群中心建置碼書；接著根據每幅影像特徵所對應的分群中心，將所有影像特徵用代碼重新進行編

3　Beis J S, Lowe D G. Shape Indexing Using Approximate Nearest-Neighbour Search in High-Dimensional Spaces[C]// Conference on Computer Vision and Pattern Recognition, Puerto Rico. 1997:1000-1006.

碼（採用分群中心頻次長條圖所對應的向量來表示），建置索引。當進行相似影像檢索時，可以使用第 1 章中提到的文字搜尋引擎所採用的方式來處理。首先將查詢影像特徵依照相同的方法進行編碼，然後搜索索引，獲得與查詢特徵編碼具有相同代碼的候選集，最後在候選集中比較相似度。這也貫徹了先定位到子空間、再進行比較的思維，進而大幅縮減了檢索時間。

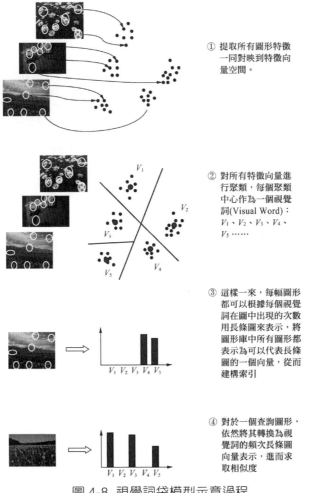

① 提取所有圖形特徵一同對映到特徵向量空間。

② 對所有特徵向量進行聚類，每個聚類中心作為一個視覺詞(Visual Word)：V_1、V_2、V_3、V_4、V_5……

③ 這樣一來，每幅圖形都可以根據每個視覺詞在圖中出現的次數用長條圖來表示，將圖形庫中所有圖形都表示為可以代表長條圖的一個向量，從而建構索引

④ 對於一個查詢圖形，依然將其轉換為視覺詞的頻次長條圖向量表示，進而求取相似度

圖 4-8 視覺詞袋模型示意過程

當面對大規模高維資料時，資料量極大會造成分群後簇內方差過大，常常沒有明顯的簇中心，進一步導致向量量化的方法存在時空效率較低的缺點。為了解決這一問題，法國 INRIA 實驗室在向量量化的基礎上提出了乘積量化（Product Quantization）的方法。乘積量化首先將 D 維向量空間平均分為 M 個子空間，然後對每個子空間進行分群形成 M 套碼書，並使用對應碼書的代碼對各子空間內的資料進行編碼。最後量化的結果使用對各子空間量化結果的笛卡爾乘積來表示，這也是稱為「乘積量化」的原因所在。

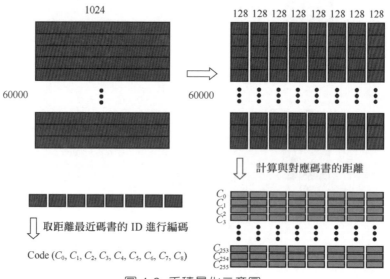

圖 4-9 乘積量化示意圖

舉個簡單的實例，假設現在有 60000 幅影像，每幅影像都使用某種相同的方式分析特徵，形成 60000 個具有 1024 維度的特徵向量，如圖 4-9 左上部分所示。首先將每個特徵向量平均分為 8 份，也就是每份 128 維，如圖 4-9 右上部分所示。然後對每一等距進行 K-means 分群，也就是每一等距都產生了 k 個分群中心。這樣就獲得 8 套碼書，每套碼書均由 k

個代碼組成。接下來，將計算 8 個等距，每個等分內的 60000 個子向量
與對應碼書的代碼間的距離，使用距離最近的代碼來量化編碼。例如假
設 k=256，那麼每套碼書都有 256 個代碼。每幅影像的特徵向量都可以
量化編碼為類似 "$Code$(28,56,32,88,252,36,77,129)" 的形式（十進位表
示），由此我們可以建置對應的索引。

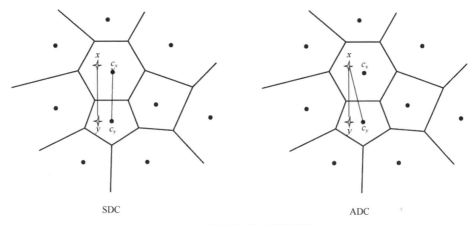

SDC   ADC

圖 4-10　兩種不同的距離計算方式

當進行相似性檢索時，仍然需要將分析的查詢向量平均分為 8 份，然後
計算每個子向量與索引中對應子空間內所有分群中心的距離。在計算距
離時有兩種方式：一種是對稱距離計算（SDC），另一種是非對稱距離
計算（ADC）。如圖 4-10 所示，SDC 在計算 x 和 y 的距離時，實際上計
算的是它們所對應代碼 C_x 和 C_y 的距離，而在 ADC 中 x 和 y 的距離計
算的是 x 和 y 所對應代碼 C_y 間的距離。也就是説，SDC 計算需要將子
向量進一步量化為代碼，SDC 實際上是計算代碼間的距離；ADC 並不
需要對子向量進行量化，計算的是查詢向量和對應索引中代碼的距離。
ADC 比 SDC 的計算代價更大，但更精確，我們一般採用 ADC 方式計算
距離。無論使用的是 ADC 還是 SDC，最後都需要將各部分獲得的距離

求和，然後進行排序，進而獲得最為相似的許多影像。在計算距離的過程中，我們雖然還是使用的線性比較方法，但比較的物件是分群演算法產生的代碼，而非所有影像特徵向量。比較次數只與分群中心的數量 k 相關，這樣就相當大地降低了比較次數和查詢時間。

隨著資料規模的變大，k 的設定值無疑也會進一步增大。雖然我們可以採用提前計算代碼間距離或查詢向量與代碼間距離的方式產生查閱資料表，但是隨著 k 值的增大，查詢時間也會顯著延長。為此，INRIA 實驗室的 Herve Jegou 將倒排索引結構引入乘積量化中來，形成 Inverted File Product Quantization（IVFPQ）[4]。如圖 4-11 所示，Herve Jegou 首先對特徵向量 Y 進行 K-means 分群，獲得 k 個粗量化（coarse quantizer）的分群中心，每個分群中心對應一個倒排列表（inverted list）。接下來計算輸入 $y \in Y$ 與其對應分群中心的殘差 $r(y)=y-q_c(y)$，將殘差 $r(y)$ 分成 m 等份，平行進行乘積量化產生量化編碼，並組合成一個 m 維的 $code$，將 Y 中所有特徵向量 y 的 id 和 $r(y)$ 的乘積量化編碼 $code$ 插入倒排列表。

當進行相似性檢索時，首先依據查詢向量與粗量化代碼的距離，直接確定相似的物件最可能出現在某個或某幾個倒排列表中。然後對查詢向量 x 進行粗量化 $q_c(x)$，計算 x 的殘差 $r(x)$ 並將其分為 m 等份，計算各部分 $r_i(x)$ 與可能存在最近鄰的倒排列表中 $code_i$ 的距離並求和，利用堆排序取得距離和最小的若干個結果。IVFPQ 克服了 PQ 必須在大範圍內進行線性比較的弱點，充分貫徹了先定位在對應子空間，然後進行線性比較的策略，大幅提高了效率。

---

4 Jegou H,Douze M,Schmid C.Product Quantization for Nearest Neighbor Search[J].IEEE Transactions on Pattern Anlysis & Machine Intelllgence,2011,33(1);117.

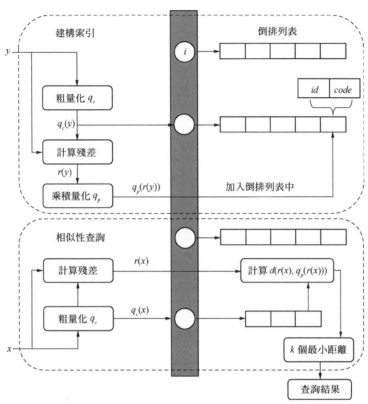

圖 4-11　倒排索引乘積量化

❸ 以局部敏感雜湊方法為基礎的索引建置與檢索

雜湊（hash）函數是一種形如 $y=h(x)$ 的對映關係，它能夠將任意長度的字元對映為固定長度的字元。通常在一般的線性串列、樹狀結構中，記錄位置和它們的關鍵字間並不存在直接的關係，而是隨機儲存的，然而雜湊函數可以在關鍵字和記錄位置之間建立起相應的對映關係。這樣我們就可以根據關鍵字直接計算出記錄所在的位置，在理想情況下，記錄的尋找時間為 $O(1)$。雜湊函數的這一優點對於解決最近鄰檢索問題提供了一個極佳的想法，但是這些普通的雜湊函數並不能保障相似的鄰近資料在經過雜湊轉換後也是鄰近的。如果我們能夠找到這樣一種雜湊函數，它能讓原本相似的資料在經過雜湊轉換後也是大機率地對映到同一個「桶」（bucket）內，讓原本不相似的資料被對映到不同的「桶」內，那麼尋找最近鄰的問題也就變得容易多了。我們只需將查詢向量進行雜湊轉換獲得對應的「桶」號，然後在這一「桶」內進行線性比較就可以了。局部敏感雜湊函數剛好就是這樣一種雜湊函數，這種雜湊函數必須滿足以下兩個條件（如圖 4-12 所示）。

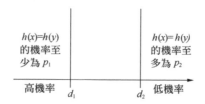

圖 4-12 局部敏感雜湊函數的條件

（1）如果 $dist=(x, y) \leq d_1$，則 $h(x)=h(y)$ 的機率至少為 p_1。

（2）如果 $dist=(x, y) \geq d_2$，則 $h(x)=h(y)$ 的機率至多為 p_2。

其中，$dist=(x, y)$ 表示 x 與 y 之間在某種度量方式下的距離，且 $d_1 < d_2$，$p_1 > p_2$，$h(x)$ 和 $h(y)$ 分別表示對 x 和 y 做雜湊轉換。滿足以上條件的雜湊函數族稱為 (d_1, d_2, p_1, p_2) 敏感的，透過一個或多個 (d_1, d_2, p_1, p_2) 敏感的雜湊函數對資料集進行雜湊轉換產生一個或多個散清單的過程稱為局部敏感雜湊函數（Locality Sensitive Hashing）。

局部敏感雜湊函數只是以一定的機率在對應的距離度量方式下是局部敏感的，脫離了實際的距離度量方式，它只是一種普通的雜湊函數。

在傑卡德相似度下最具代表性的局部敏感雜湊函數是 MinHash。MinHash 由 Andrei Broder 在 1997 年為了解決重複網頁檢測和大規模分群問題而提出[5]。MinHash 將求取兩個集合間傑卡德相似度的問題轉化為兩個最小雜湊函數值相等的機率問題。經簡單推導可知，兩個集合 A 和 B 的元素經許多次的行隨機排列轉換後，取得的最小雜湊值 $h_{min}(A)$、$h_{min}(B)$ 相等的機率與 A 和 B 的傑卡德相似度等同。也就是說，可以透過計算 $h_{min}(A)$ 等於 $h_{min}(B)$ 的機率來取得 A 與 B 的相似度。在實際應用中，由於對大規模資料的隨機排列轉換會消耗極大的時間和運算資源，所以這樣做是不實際的。研究人員找到一種使用許多雜湊函數來模擬行隨機排列轉換的方法，例如對於行數為 rows 的特徵矩陣，形如 $h(r)=(r+1)$ mod *rows* 的雜湊函數就可以根據目前行號 r 計算出行排列轉換後的行號 $h(r)$。利用許多這樣的雜湊函數進行行號的再次對映，必然能實現特徵矩陣的隨機行排列轉換。將特徵矩陣的行號進行上述雜湊轉換後依然取最小行號，進而產生最小雜湊簽名矩陣，在該矩陣中統計集合 A 和 B 對應項相等的機率。由於傑卡德相似度 $JAC(x, y)=1-dist(x, y)$，如果 $dist(x, y) \leq d_1$，那麼 $JAC(x, y) \geq 1-d_1$，而 x 與 y 的傑卡德相似度 $JAC(x, y)$ 又等於最小雜湊函數對 x、y 對映後結果相等的機率，所以 MinHash 函數族是 $(d_1, d_2, 1-d_1, 1-d_2)$ 敏感的。

在漢明距離下最具代表性的局部敏感雜湊函數是隨機位元取樣（Random bits sampling）$h(y)=y_k$，其中 $k \in \{1, 2, \cdots, d\}$，$y$ 是一個每一維的設定值為 0 或 1 的二進位向量。隨機位取樣的基本思維是隨機選擇 d 維特徵向量的某一維，這樣一來，兩個集合 A 和 B 經隨機位取樣後相等的機率便等於 A 和 B 的相似度。隨機位元取樣雜湊函數族是 $(d_1, d_2, 1-d_1/d, 1-d_2/d)$ 敏感的。

5　Broder A. On the resemblance and containment of documents[C]// sequences. IEEE Computer Society, 1997:21.

在餘弦距離下最具代表性的局部敏感雜湊函數是隨機投影（Random projection）。隨機投影將向量 x 和一個由法向量 r 定義的隨機超平面做點積，將點積的符號作為雜湊函數的輸出，即 $h(x)=\text{sgn}(x.r)=\text{sgn}(r^{\text{T}}x)=\pm 1$。隨機超平面將空間分為兩部分，而 $h(x)$ 的設定值取決於向量 x 在隨機超平面的哪一邊。通常將法向量 r 所在的空間稱為正空間，將另一半稱為負空間。我們可以根據向量與法向量間的夾角來判斷該向量所在的空間，與法向量的夾角是銳角的向量在超平面的正空間，與法向量的夾角是鈍角的向量在超平面的負空間。如圖 4-13 所示，法向量 r_1 和 r_2 分別定義了一個超平面，圖中分別用不同樣式的虛線來表示。向量 x 和 y 分別在 r_1 定義的超平面的兩邊，$h(x)$ 與 $h(y)$ 的符號各異；然而向量 x 和 y 在 r_2 定義的超平面的同一邊，$h(x)$ 與 $h(y)$ 的符號相同。由於法向量 r 是隨機的，我們可以進一步確定兩個向量 x、y 被分在超平面同一邊的機率，也就是 $h(x)=h(y)$ 的機率是 $1-\dfrac{\theta}{\pi}$，所以該敏感雜湊函數是 $(d_1, d_2, 1-d_1/180, 1-d_2/180)$ 的。

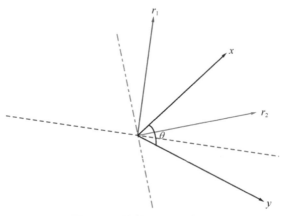

圖 4-13 隨機投影示意圖

在歐氏距離下，最具代表性的局部敏感雜湊函數也是利用隨機投影原理建置的。它形如 $h(x)=\left|\dfrac{x \cdot r+b}{w}\right|$，其中 r 是一個隨機向量，w 是桶寬，b 是

一個在 [0, a] 之間均勻分佈的隨機變數,而 $x \cdot r$ 可以看作將向量 x 向 r 上做投影的操作。也就是將原始資料空間中的資料 x 投影到一條由被均分為寬度為 w 的許多線段組成的隨機直線上,寬度為 w 的線段可以看作一系列的「雜湊桶」,如圖 4-14 所示。在原空間歐式距離相近的資料會具有極高的可能性被投影到同一「桶」中,而這一可能性取決於兩個資料點(x 和 y)間形成的連線(長度為 d)與隨機直線形成的夾角 θ。例如當 θ=90° 時,這一連線與隨機直線垂直,x 和 y 必定落入同一「桶」中。假設 $d \leq a/2$,那麼至少有 50% 的機率被投影到同一桶中;當 $d \geq 2a$ 時,若要點被投影到同一桶中,$\cos\theta$ 最多為 1/2,那麼 θ 必須在 60°~90°,很容易得出其發生的機率為 1/3。由此可以得出該敏感雜湊函數是 $(a/2, 2a, 1/2, 1/3)$ 敏感的。

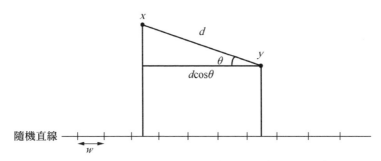

圖 4-14　歐式距離下的局部敏感雜湊函數原理示意圖

透過使用以上介紹的局部敏感雜湊函數,我們現在已經能夠大機率將原本相鄰的資料對映到同一個桶中,將不相鄰的資料對映到不同的桶中。但這只是大機率而非必然,所以就會存在將原本相鄰的資料對映到不同桶中,將原本不相鄰的資料對映到同一桶中的情況。為了加強正確對映的機率,我們可以透過以下兩種策略來進一步最佳化局部敏感雜湊的效果:在一個散清單中使用更多的局部敏感雜湊函數;使用更多的雜湊表。

這兩種策略又衍生出以下具體的方法。

（1）建立多個獨立的雜湊表，每個散清單又由 k 個局部敏感雜湊函數產生。

（2）「與」操作使用 k 個局部敏感雜湊函數，且只有當兩個資料 x 和 y 的這 k 個局部敏感雜湊函數的雜湊值都相等的條件下，才會將資料 x 和 y 對映到同一桶中。

（3）「或」操作同樣適用 k 個局部敏感雜湊函數，而當兩個資料 x、y 的這 k 個局部敏感雜湊函數的雜湊值至少有一對相同時，資料 x、y 就會被對映到同一桶中。

（4）「與」和「或」操作串聯使用。

以局部敏感雜湊進行資料索引和檢索為基礎的方法同樣採用離線建置索引 -- 線上檢索的方式進行。首先選取某個相似度或某種距離度量方式下滿足 (d_1, d_2, p_1, p_2) 敏感的局部敏感雜湊函數族；然後根據對檢索結果準確率的要求，確定雜湊表的個數 L 以及每個散清單中局部敏感雜湊函數的數量 K；最後將所有資料使用 K 個局部敏感雜湊函數對映到對應的桶中，進而形成 L 個雜湊表對資料進行離線儲存。在進行資料檢索時，利用局部敏感雜湊函數的對映值取得對應的桶，並計算查詢資料與桶內資料之間的相似度或距離，根據相似度或距離的排序傳回許多查詢資料的近似最近鄰。

下面將透過一個隨機投影局部敏感雜湊函數的實作方式，來進一步了解局部敏感雜湊函數，以及以 LSH 為基礎的資料索引建置與檢索。

⧗ 程式 4-10：隨機投影 LSH

```
package com.ai.deepsearch.index;

import java.util. *;
```

```java
/**
 * LSH 索引與檢索
 */
public class LSH {
    // 求點積
    private double dotProduct(double[] v1, double[] v2) {
        int dimension = v1.length;
        double product = 0;
        for (int i = 0; i < dimension; i++) {
            product += v1[i] * v2[i];
        }
        return product;
    }


    // 隨機投影方式產生雜湊
    private int generateHash(double[] vector) {
        int dimension = vector.length;
        Random random = new Random();

        // 產生隨機超平面
        double[] randHyperPlane = new double[dimension];

        for (int i = 0; i < dimension; i++) {
            randHyperPlane[i] = random.nextGaussian();
        }

        return dotProduct(vector, randHyperPlane) > 0 ? 1 : 0;
    }

    // 合併多個雜湊值，提高效率
    private int combineHashes(int hashSize, double[] vector) {
        int[] hashes = new int[hashSize];
```

```java
    for (int i = 0; i < hashSize; i++) {
        hashes[i] = generateHash(vector);
    }


    int combine = 0;
    int power = 1;
    for (int i = 0; i < hashes.length; i++) {
        combine += hashes[i] == 0 ? 0 : power;
        power *= 2;
    }


    return combine;
}


// 雜湊桶類別
private class Bucket {
    private ArrayList<double[]> data;

    public Bucket() {
        this.data = new ArrayList<double[]>();
    }

    public void add(double[] vector) {
        data.add(vector);
    }

    public ArrayList<double[]> getData() {
        return this.data;
    }
}

// 雜湊表類別
```

```java
private class HashTable {
    private HashMap<Integer, Bucket> buckets;

    private int hashSize = 0;

    public HashTable(int hashSize) {
        this.buckets = new HashMap<Integer, Bucket>();
        this.hashSize = hashSize;
    }

    public void add(double[] vector) {
        int key = combineHashes(hashSize, vector);
        if (buckets.containsKey(key)) {
            buckets.get(key).add(vector);
        } else {
            Bucket bucket = new Bucket();
            bucket.add(vector);
            buckets.put(key, bucket);
        }
    }

    public Bucket query(double[] qVector) {
        int key = combineHashes(hashSize, qVector);
        if (buckets.containsKey(key)) {
            return buckets.get(key);
        } else {
            return null;
        }
    }

    public HashMap<Integer, Bucket> getBuckets() {
        return buckets;
```

```
        }
    }


    // 雜湊表個數 L
    private int tablesSize = 0;
    // 每個散清單中雜湊函數的個數 K
    private int hashSize = 0;


    private ArrayList<HashTable> tables;


    public LSH(int tableSize, int hashSize) {
        this.tablesSize = tableSize;
        this.hashSize = hashSize;


        this.tables = new ArrayList<HashTable>();
        for (int i = 0; i < tablesSize; i++) {
            HashTable table = new HashTable(hashSize);
            tables.add(table);
        }
    }


    public int getNumOfTables() {
        return this.tablesSize;
    }


    public int getNumOfHashes() {
        return this.hashSize;
    }


    public ArrayList<HashTable> getHashTables() {
        return this.tables;
    }
```

```java
// 建置索引
public void index(double[] vector) {
    for (HashTable table : tables) {
        table.add(vector);
    }
}


// 餘弦距離
public double cosineDistance(double[] v1, double[] v2) {
    double distance = 0;
    distance = 1 - dotProduct(v1, v2) / Math.sqrt(dotProduct(v1,
v1) * dotProduct(v2, v2));
    return distance;
}


// 查詢近似最近鄰
public List<double[]> queryNeighbours(final double[] qVector, int
count) {
    Set<double[]> neighbourSet = new HashSet<double[]>();
    for (HashTable table : tables) {
        Bucket bucket = table.query(qVector);
        if (bucket == null) {
            return null;
        } else {
            List<double[]> data = bucket.getData();
            neighbourSet.addAll(data);
        }
    }

    List<double[]> neighbours = new ArrayList<double[]>(neighbourSet);
    Collections.sort(neighbours, new Comparator<double[]>() {
```

```java
        @Override
        public int compare(double[] v1, double[] v2) {
            Double v1Dis = cosineDistance(qVector, v1);
            Double v2Dis = cosineDistance(qVector, v2);
            return v1Dis.compareTo(v2Dis);
        }
    });

    if (neighbours.size() > count) {
        neighbours.subList(0, count);
    }

    return neighbours;
}

public static void main(String args[]) {

    double[][] indexData = {{1}, {3}, {4}, {7}, {8}, {9}, {11}};
    double[] queryData = {2};

    LSH lsh = new LSH(1, 4);

    // 索引階段
    System.out.println("#############\n        索引開始
\n#############");
    for (int i = 0; i < indexData.length; i++) {
        lsh.index(indexData[i]);
    }

    ArrayList<HashTable> tables = lsh.getHashTables();
    System.out.println(" 雜湊表數 :" + tables.size());
```

```java
        for (int i = 0; i < tables.size(); i++) {
            HashTable table = tables.get(i);
            HashMap<Integer, Bucket> buckets = table.getBuckets();
            for (Integer key : buckets.keySet()) {
                Bucket bucket = buckets.get(key);
                ArrayList<double[]> data = bucket.getData();
                System.out.print("桶號 " + key + ":");
                for (int j = 0; j < data.size(); j++) {
                    double[] elemData = data.get(j);
                    for (int k = 0; k < elemData.length; k++) {
                        System.out.print(elemData[k]);
                        if (k != elemData.length - 1) System.out.
print(",");
                    }
                    if (j != data.size() - 1) System.out.print("-->");
                }
                System.out.println();
            }
        }

        // 檢索階段
        System.out.println("#############\n        檢索開始
\n#############");
        List<double[]> result = lsh.queryNeighbours(queryData, 1);
        if (result == null) {
            System.out.print("未找到近似最近鄰!");
        } else {
            System.out.print("檢索結果:");
            for (int i = 0; i < result.size(); i++) {
                double[] ann = result.get(i);
                for (int j = 0; j < ann.length; j++) {
                    System.out.print(ann[j]);
```

```
                    if (j != ann.length - 1) System.out.print(",");
            }
            if (i != result.size() - 1) System.out.print("-->");
        }
    }
}
```

4.5 本章小結

利用分析到的影像特徵高效率地建置索引以加快檢索處理程序,進一步
實現可實際應用的影像搜尋引擎,還需要解決哪些技術問題呢?

本章從影像特徵降維處理講起,介紹了影像特徵標準化以及影像特徵相
似度的度量方法,分析了以樹結構、向量量化、局部敏感雜湊為基礎的
三種索引建置與檢索方法,並透過實作方式呈現一個隨機投影局部敏感
雜湊函數的程式,讓讀者更深刻地了解局部敏感雜湊函數的理論和方
法。

建置一個以深度學習為基礎的 Web 影像搜尋引擎

在第 2 ～ 4 章，我們分別介紹了影像特徵分析的傳統方法以及以深度學習為基礎的方法、影像特徵的索引和檢索方法，由此便掌握了建置一個影像搜尋引擎的基本理論和方法。本章，將説明如何從零開始逐步建置一個以深度學習為基礎的 Web 影像搜尋引擎。

5.1 架構分析與技術路線

5.1.1 架構分析

一個 Web 系統必然會有前後端之分，影像搜尋引擎也不例外。首先我們需要一組能夠接收使用者傳送影像資訊，並即時將查詢結果回饋給使用者的前端頁面，另外還需要一個能夠分析影像特徵，並與特徵索引函數庫進行比較傳回比較結果的後端系統，然後透過某種 Web 架構將前後端連接起來，如圖 5-1 所示。此外，我們還需要一個能夠對影像函數庫內一個一個影像進行特徵分析，並形成影像特徵索引函數庫的工具。

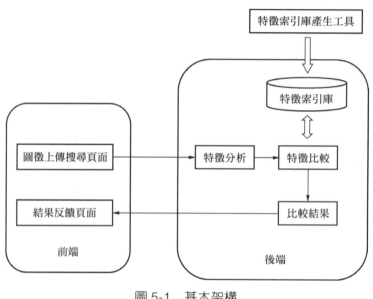

圖 5-1 基本架構

5.1.2 技術路線

由於本書中的程式均採用 Java 語言實現，為了保持一致的風格與讀者體驗，本章將繼續沿用 Java。前端採用經典的 HTML+CSS+JQuery 方式實現，後端影像特徵分析部分將基於 DeepLearning4J 實現。由於 Java Web 領域存在大量設定繁雜的各種架構，如果使用它們會本末倒置，嚴重干擾本章的主題。為了讓讀者能夠更清晰地了解建置一個以 Web 為基礎的影像搜尋引擎的基本原理，本章將使用 Java Servlet 標準將前端和後端連接起來。同樣為突出特徵索引函數庫產生工具基本原理的呈現，該工具將採用包裝為 Jar 命令列程式的形式。

5.2 程式實現

5.2.1 開發環境架設

源於 Java 卓越的跨平台性，在 Windows、Linux、MAC 系統中均可以架設對應的開發環境。對於 IDE，選擇 Eclipse、IntelliJ IDEA 或是 Netbeans 均可。由於該專案影像特徵分析部分基於 DeepLearning4J 實現，而其又依賴大量的 Jar 套件，所以需要引用 Maven 來對依賴進行管理和自動化建置。IntelliJ IDEA 內建 Maven，並且是 DeepLearning4J 官方推薦的 IDE，本章將使用它來開發此專案。但是因為目前版本的 DeepLearning4J 需要 Maven3.2.5 以上版本的支援，所以需要對應版本的 IntelliJ IDEA，在此推薦到 JetBrains 官方網站中下載安裝最新版，如圖 5-2 所示。IntelliJ IDEA 的社區（Community）版就可以滿足本專案的需求。

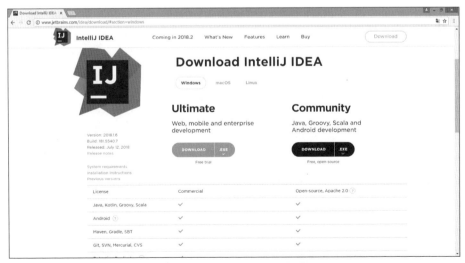

圖 5-2 IntelliJ IDEA 社區版

5.2.2 專案實現

在該專案下，需要分別建立影像搜尋引擎和特徵索引函數庫產生工具兩個子專案，下面來分別說明。

1 特徵索引函數庫產生工具子專案

首先開啟 IntelliJ IDEA 建立新專案（"Create New Project"），然後在新專案（"New Project"）對話方塊 "Maven" 項下選擇從原型建立（"Create from archetype"），並從中選擇 "org.apache. maven.archetypes:maven-archetype-quickstart" 原型範本建立一個簡單的 Java 本機應用，如圖 5-3 所示。

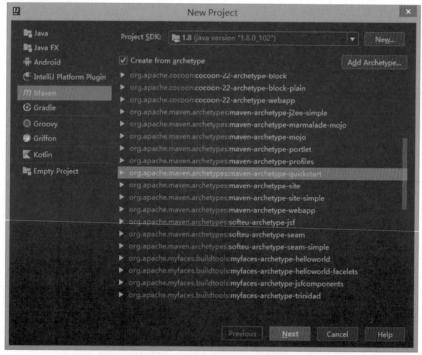

圖 5-3 透過 Maven quickstart 原型範本建立專案

接下來，如圖 5-4 所示，填入對應的 "GroupId" 和 "ArtifactId"。GroupId 和 ArtifactId 標識了 Maven 專案的唯一性。其中 GroupId 又分為多個段，第一段為域可以是 org、com、cn 等，第二段為公司名，第三段為專案名稱，而 Artifact 表示功能模組名稱。此專案中，我們在 GroupId 中填入 "com.ai.deepsearch"，在 Artifact 中填入 "GenerateImgsFeatDBTool"，它的全標識就是 "com.ai.deepsearch.GenerateImgsFeatDBTool"。

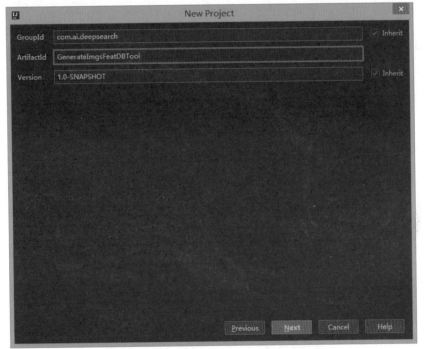

圖 5-4 填寫對應的 Maven 專案標識

下一步，如圖 5-5 所示，IDE 會顯示 Maven 的路徑選擇、使用者設定檔、倉庫設定等資訊，在此我們使用 IntelliJ IDEA 內建的 Maven 即可。點擊 "Next" 核對專案名稱，選擇對應的專案儲存路徑，點擊 "Finish" 後，Maven 將自動完成該專案基本結構的建立。

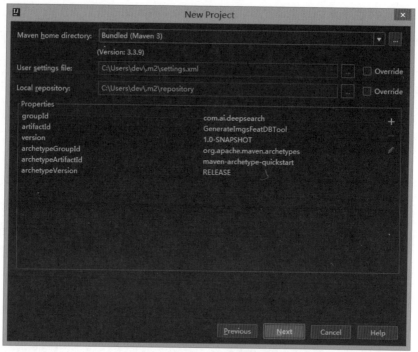

圖 5-5　Maven 基本資訊設定

經過一段時間的等待，Maven 成功地下載完成各種所需的外掛程式，並依據 maven-archetype- quickstart 原型範本建立專案的基本結構，如圖 5-6 所示。可以看到，src、main、test、java 等資料夾，以及 com. ai.deepsearch 套件，及其下的 APP 和 APPTest 類別等 Java 本機專案常用的結構和檔案均已建立。

圖 5-6　Maven quickstart 原型範本自動建立的專案結構

接下來將編輯 Maven 專案的描述檔案 pom.xml（專案物件模型），引用
該專案所需的各種依賴函數庫。下面我們先把編輯好的 pom.xml 檔案列
出，程式如下，後面再詳細解釋。

⏳ 程式 5-1：特徵索引函數庫產生工具子專案 pom.xml

```xml
<?xml version="1.0" encoding="UTF-8"?>

<projectxmlns="http://maven.apache.org/POM/4.0.0" xmlns:xsi="http://
www.w3.org/2001/XMLSchema-instance"
  xsi:schemaLocation="http://maven.apache.org/POM/4.0.0
http://maven.apache.org/xsd/maven-4.0.0.xsd">
  <modelVersion>4.0.0</modelVersion>

  <groupId>com.ai.deepsearch</groupId>
  <artifactId>GenerateImgsFeatDBTool</artifactId>
  <version>1.0-SNAPSHOT</version>
```

```xml
<name>GenerateImgsFeatDBTool</name>

<properties>
  <project.build.sourceEncoding>UTF-8</project.build.sourceEncoding>
  <maven.compiler.source>1.7</maven.compiler.source>
  <maven.compiler.target>1.7</maven.compiler.target>
  <!-- Change the nd4j.backend property to nd4j-cuda-8.0-platform to
use CUDA GPUs -->
  <nd4j.backend>nd4j-native-platform</nd4j.backend>
  <java.version>1.8</java.version>
  <nd4j.version>0.9.1</nd4j.version>
  <dl4j.version>0.9.1</dl4j.version>
  <datavec.version>0.9.1</datavec.version>
  <arbiter.version>0.9.1</arbiter.version>
  <logback.version>1.1.7</logback.version>
  <mapdb.version>3.0.6</mapdb.version>
  <commons-cli.version>1.4</commons-cli.version>
</properties>

<dependencyManagement>
  <dependencies>
    <dependency>
      <groupId>org.nd4j</groupId>
      <artifactId>nd4j-native-platform</artifactId>
      <version>${nd4j.version}</version>
    </dependency>
    <dependency>
      <groupId>org.nd4j</groupId>
      <artifactId>nd4j-cuda-7.5-platform</artifactId>
      <version>${nd4j.version}</version>
    </dependency>
    <dependency>
```

```xml
      <groupId>org.nd4j</groupId>
      <artifactId>nd4j-cuda-8.0-platform</artifactId>
      <version>${nd4j.version}</version>
    </dependency>
  </dependencies>
</dependencyManagement>

<dependencies>
  <dependency>
    <groupId>junit</groupId>
    <artifactId>junit</artifactId>
    <version>4.11</version>
    <scope>test</scope>
  </dependency>
  <dependency>
    <groupId>org.nd4j</groupId>
    <artifactId>${nd4j.backend}</artifactId>
  </dependency>
  <dependency>
    <groupId>org.deeplearning4j</groupId>
    <artifactId>deeplearning4j-core</artifactId>
    <version>${dl4j.version}</version>
  </dependency>
  <dependency>
    <groupId>org.deeplearning4j</groupId>
    <artifactId>arbiter-deeplearning4j</artifactId>
    <version>${arbiter.version}</version>
  </dependency>
  <dependency>
    <groupId>ch.qos.logback</groupId>
    <artifactId>logback-classic</artifactId>
    <version>${logback.version}</version>
```

```
    </dependency>
    <dependency>
      <groupId>org.mapdb</groupId>
      <artifactId>mapdb</artifactId>
      <version>${mapdb.version}</version>
    </dependency>
    <dependency>
      <groupId>commons-cli</groupId>
      <artifactId>commons-cli</artifactId>
      <version>${commons-cli.version}</version>
    </dependency>
  </dependencies>

  <build>
    <pluginManagement><!-- lock down plugins versions to avoid using
Maven defaults
(may be moved to parent pom) -->
      <plugins>
        <plugin>
          <artifactId>maven-clean-plugin</artifactId>
          <version>3.0.0</version>
        </plugin>
        <!-- see http://maven.apache.org/ref/current/maven-core/
default-bindings.html#Plugin_bindings_for_jar_packaging -->
        <plugin>
          <artifactId>maven-resources-plugin</artifactId>
          <version>3.0.2</version>
        </plugin>
        <plugin>
          <artifactId>maven-compiler-plugin</artifactId>
          <version>3.7.0</version>
        </plugin>
```

```xml
    <plugin>
      <artifactId>maven-surefire-plugin</artifactId>
      <version>2.20.1</version>
    </plugin>
    <plugin>
      <artifactId>maven-jar-plugin</artifactId>
      <version>3.0.2</version>
    </plugin>
    <plugin>
      <artifactId>maven-install-plugin</artifactId>
      <version>2.5.2</version>
    </plugin>
    <plugin>
      <artifactId>maven-deploy-plugin</artifactId>
      <version>2.8.2</version>
    </plugin>
    </plugins>
  </pluginManagement>
  </build>
</project>
```

可以看到 pom.xml 檔案大致分為幾個段落。首先是專案的基礎資訊 groupId、artifactId、version，以及包裝方式 packaging 和專案名稱 name，這些都是根據前面填入的專案資訊和 Maven 原型範本自動產生的。接下來 properties 部分定義了編譯器的版本、nd4j 函數庫的後端選擇 -- nd4j.backend 以及所有需要引用依賴的版本。下面的 dependencyManagement 部分對 nd4j 函數庫所需要的兩種後端 nd4j-native-platform 和 nd4j-cuda-8.0-platform 進行了宣告，便於我們在使用時根據需要進行引用。dependencies 部分定義了需要引用的各種依賴：nd4j 的後端（我們這裡只使用 CPU 而不使用 GPU，所以引用的是 nd4j-native-platform)、DeepLearning4J 的 核 心 元 件 deeplearning4j-core、DeepLearning4J 中用於超參數最佳化的 arbiter-deeplearning4j、DeepLearning4J 中用於記錄檔系統的 logback、嵌入式資料庫 mapdb、用於對命令列參數進行解析的 apache commons-cli 函數庫。最後的 build 部分中定義了最後產生的專案名稱，以及各種用於編譯、包裝、安裝、部署、清理、資源的 Maven 外掛程式。

在編輯並儲存了 pom.xml 檔案之後，IntelliJ IDEA 會提示是否自動引用依賴，我們選擇自動引用，IDE 會立即下載相關的函數庫檔案。當所有的依賴函數庫下載完成以後，就可以編輯程式了。首先將 Maven 自動產生的 App 等不需要的類別刪除掉，然後新增一個類別 GenerateImgsFeatDBTool，程式如下。

▌ 程式 5-2：GenerateImgsFeatDBTool.java

```java
package com.ai.deepsearch;

import org.apache.commons.cli.*;
import org.datavec.image.loader.NativeImageLoader;
import org.deeplearning4j.nn.graph.ComputationGraph;
import org.deeplearning4j.util.ModelSerializer;
import org.mapdb.DB;
import org.mapdb.DBMaker;
import org.mapdb.Serializer;
import org.nd4j.linalg.api.ndarray.INDArray;
import org.nd4j.linalg.dataset.api.preprocessor.DataNormalization;
import org.nd4j.linalg.dataset.api.preprocessor.VGG16ImagePreProcessor;

import java.io.File;
import java.io.IOException;
import java.util.Map;
```

```java
import java.util.concurrent.ConcurrentMap;

/**
 * 影像特徵函數庫產生工具
 */
public class GenerateImgsFeatDBTool {
    private DB db;
    private ConcurrentMap<String, double[]> map;
    private static ComputationGraph vgg16Model;

    // 建立一個 mapdb 資料庫
    private void initDB(String dbName) {
        System.out.println("GenerateImgsFeatDBTool init db");
        db = DBMaker.fileDB(dbName).fileMmapEnable().make();
        map = db.hashMap("feat_map", Serializer.STRING, Serializer.
DOUBLE_ARRAY).createOrOpen();
    }

    // 載入預訓練的 VGG16 模型檔案
    private void loadVGGModel(String modelFilePath) throws IOException {
        File vgg16ModelFile = new File(modelFilePath);
        vgg16Model = ModelSerializer.restoreComputationGraph
(vgg16ModelFile);
    }

    // 使用預訓練 VGG16 模型獲得影像的 FC2 層特徵
    private INDArray getImgFeature(File imgFile) throws IOException {
        NativeImageLoader loader = new NativeImageLoader(224, 224, 3);
        INDArray imageArray = loader.asMatrix(imgFile);
        DataNormalization scaler = new VGG16ImagePreProcessor();
        scaler.transform(imageArray);
        Map<String, INDArray> map = vgg16Model.feedForward(imageArray,
```

```
false);
        INDArray feature = map.get("fc2");
        return feature;
    }


    // 陣列類型轉換函數
    private double[] INDArray2DoubleArray(INDArray indArr) {
        String indArrStr = indArr.toString().replace("[", "").replace
("]", "");
        String[] strArr = indArrStr.split(",");
        int len = strArr.length;
        double[] doubleArr = new double[len];
        for (int i = 0; i < len; i++) {
            doubleArr[i] = Double.parseDouble(strArr[i]);
        }
        return doubleArr;
    }


    // 將一個影像檔夾內的影像轉為特徵碼並存入特徵資料庫中
    private void exportImgsFeature2DB(String imgDirName) throws
IOException {
        File dir = new File(imgDirName);
        if (dir.isDirectory()) {
            File[] fileList = dir.listFiles();
            int len = fileList.length;
            for (int i = 0; i < len; i++) {
                INDArray feat = getImgFeature(fileList[i]);
                double[] featD = INDArray2DoubleArray(feat);
                String imgName = fileList[i].getName();
                System.out.println("GenerateImgsFeatDBTool:" + i + ":"
+ imgName + "," + featD.toString());
                map.put(imgName, featD);
```

```
            }
        } else {
            System.out.println(imgDirName + "is not a directory!");
        }
        db.close();
        System.out.println("GenerateImgsFeatDBTool close db");
    }

    public static void main(String[] args) {
        String usage = "java -jar GenerateImgsFeatDBTool.jar [-h] -m
模型路徑名稱 -d 特徵函數庫路徑名稱 -i 影像檔夾路徑名稱 ";
        HelpFormatter formatter = new HelpFormatter();
        formatter.setWidth(200);
        CommandLineParser parser = new DefaultParser();


        Option help = new Option("h",false," 顯示說明資訊 ");
        Option model=Option.builder("m").hasArg().argName("model").desc
(" 模型路徑名稱 ")
.build();
        Option database = Option.builder("d").hasArg().argName
("database").desc(" 影像特徵函數庫路徑名稱 ").build();
        Option img = Option.builder("i").hasArg().argName("imgdir").
desc(" 用於建置特徵函數庫的影像檔夾路徑全名 ").build();


        Options options = new Options();
        options.addOption(help);
        options.addOption(model);
        options.addOption(database);
        options.addOption(img);


        String modelFilePath=null;
        String dbFilePath = null;
```

```java
        String imgsDir = null;

    try {
        CommandLine line = parser.parse(options, args);
        if (line.getOptions().length > 0) {
            if (line.hasOption("h")) {
                formatter.printHelp(usage, options);
            }
            if (line.hasOption("m")) {
                modelFilePath = line.getOptionValue("m");
            }
            if((!line.hasOption("h"))&&(!line.hasOption("m"))) {
                System.out.println(" 缺少模型參數 !");
            }
            if (line.hasOption("d")) {
                dbFilePath = line.getOptionValue("d");
            }
            if((!line.hasOption("h"))&&(!line.hasOption("d"))) {
                System.out.println(" 缺少特徵函數庫參數 !");
            }
            if (line.hasOption("i")) {
                imgsDir = line.getOptionValue("i");
            }
            if((!line.hasOption("h"))&&(!line.hasOption("i"))) {
                System.out.println(" 缺少影像檔夾參數 !");
            }
        } else {
            System.out.println(" 參數為空 !");
        }
    } catch (ParseException e) {
        String message=e.getMessage();
        String[] messages=message.split(":");
```

```java
        if("Missing argument for option".equals(messages[0])) {
            if("m".equals(messages[1].trim())) {
                System.out.println(" 缺少模型參數值 !");
            }
            if("d".equals(messages[1].trim())) {
                System.out.println(" 缺少特徵函數庫參數值 !");
            }
            if("".equals(messages[1].trim())) {
                System.out.println(" 缺少影像檔夾參數值 !");
            }
        }
    }


    if ((modelFilePath != null)&&(dbFilePath != null) && (imgsDir
!= null)) {
        GenerateImgsFeatDBTool tool = new GenerateImgsFeatDBTool();
        tool.initDB(dbFilePath);
        try {
            tool.loadVGGModel(modelFilePath);
            tool.exportImgsFeature2DB(imgsDir);
        } catch (IOException e) {
            e.printStackTrace();
            System.out.println("GenerateImgsFeatDBTool 遇到錯誤 " +
e.getMessage());
        }
    }
  }
}
```

GenerateImgsFeatDBTool 類別透過載入預訓練的 VGG16 模型，分析影像檔夾內每幅影像在 FC2 層上形成的特徵，並將這些特徵存入事先建立的嵌入式資料庫 mapdb 中。這些功能又分解為 initDB（建立 mapdb 資料庫）、loadVGGModel（載入預訓練的 VGG16 模型）、getImgFeature（使用預訓練 VGG16 模型獲得影像的 FC2 層特徵）、INDArray2DoubleArray（陣列類型轉換）、exportImgsFeature2DB（將影像檔夾內的影像轉為特徵碼，並存入特徵資料庫中）5 個函數。在 main 函數中，我們利用 apache commons-cli 函數庫來解析命令列參數，進而建立命令列程式。

在 GenerateImgsFeatDBTool 類別建立完成之後，就要對它進行編譯並包裝為 jar 形式的命令列程式。為了使包裝後的程式能夠獨立使用，需要將所有引用的依賴函數庫都包裝到 jar 中，因此要使用 Maven 的 assembly 外掛程式。首先點擊選單，選擇 "View → Tool Windows → Maven Projects"，使 IntelliJ IDEA 顯示 Maven Projects 管理介面。但是它裡面並沒有我們所需要的 assembly，這時需要點擊選單，選擇 "Run → Edit Configurations" 進行設定。

如圖 5-7 所示，點擊左上角的綠色加號，在出現的 Add New Configuration 中選擇 "Maven"，在其後出現的對話方塊（如圖 5-8 所示）中填入對應的 Name 為 "assembly"，Parameters 頁簽項下的 Command line 為 "assembly:assembly"。點擊 "OK" 後，會發現在 Maven Projects 介面下多出了一個 Run Configurations 項，在它的下面有一個名為 assembly 的鋸齒項。此外，我們還需要在 pom.xml 檔案 "build → pluginManagement → plugins" 下，加入 assembly 外掛程式的相關設定，見程式 5-3。

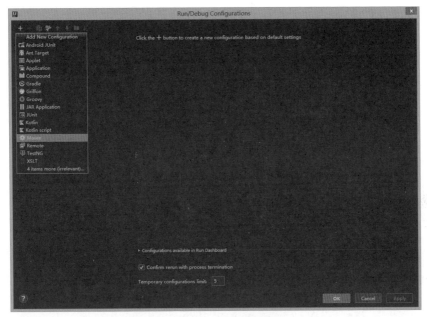

圖 5-7 Run Configurations

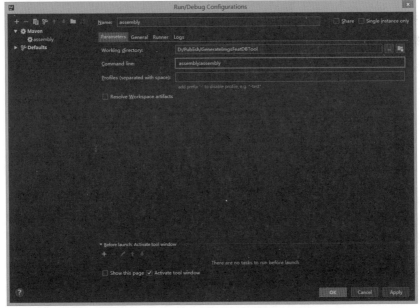

圖 5-8 在 Run Configurations 中設定 assembly 指令

⧗ **程式 5-3：assembly 外掛程式的設定**

```
<plugin>
  <artifactId>maven-assembly-plugin</artifactId>
  <configuration>
    <archive>
      <manifest>
        <mainClass>com.ai.deepsearch.GenerateImgsFeatDBTool</mainClass>
      </manifest>
    </archive>
    <descriptorRefs>
      <descriptorRef>jar-with-dependencies</descriptorRef>
    </descriptorRefs>
  </configuration>
</plugin>
```

這樣一來，我們就可以透過 Maven Projects 介面下新建立的 assembly 項來產生命令列 jar。點擊選單，依次選擇 "Run → Run assembly"，IntelliJ IDEA 會下載對應的外掛程式，並將各種依賴函數庫和編譯好的 GenerateImgsFeatDBTool.class 檔案，統一在專案 target 資料夾下包裝成可執行的 jar 套件。

最後在命令列狀態下執行 java -jar GenerateImgsFeatDBTool.jar –h，系統傳回資訊如圖 5-9 所示。

圖 5-9 執行 GenerateImgsFeatDBTool.jar

2 影像搜尋引擎子專案

影像搜尋引擎子專案與特徵索引函數庫產生工具子專案相同，都需要使用 Maven 建立。採用同樣的方法建立新專案 "Create New Project"，然後在新專案 "New Project" 對話方塊 Maven 項下選擇從原型建立（"Create from archetype"），並從中選擇 "org.apache.maven. archetypes:maven-archetype-webapp" 原型範本建立一個簡單的 Java Web 應用。點擊 "Next"，在對話方塊 GroupId 處填入 "com.ai.deepsearch"，在 ArtifactId 處填入 "ImageSearchEngine"，該子專案的全標識名稱為 com.ai.deepsearch.ImageSearchEngine。繼續點擊 "Next"，IDE 會顯示 Maven 的路徑選擇、使用者設定檔、倉庫設定等資訊。在此不用修改任何資訊，點擊 "Next"，核心對專案名稱，選擇對應的專案儲存路徑點擊 "Finish" 後，Maven 將自動完成該專案基本結構的建立。

經過一段時間的等待，Maven 成功地下載完成各種所需的外掛程式，並依據 maven-archetype-webapp 原型範本建立專案的基本結構，如圖 5-10 所示。可以看到 src、main、webapp、WEB-INF 資料夾，以及 WEB-INF 下的 web.xml 和 webapp 下的 index.jsp 檔案等 Java Web 專案下常用的結構和檔案均已建立。

圖 5-10　Maven webapp 自動建立的專案結構

下面我們依然要編輯該子專案的物件模型描述檔案 pom.xml，引用該專案所需的各種依賴函數庫，程式如下。

程式 5-4：影像搜尋引擎子專案 pom.xml

```xml
<?xml version="1.0" encoding="UTF-8"?>

<project xmlns="http://maven.apache.org/POM/4.0.0"
xmlns:xsi="http://www.w3.org/2001/XMLSchema-instance"
  xsi:schemaLocation="http://maven.apache.org/POM/4.0.0
http://maven.apache.org/xsd/maven-4.0.0.xsd">
```

```xml
<modelVersion>4.0.0</modelVersion>

<groupId>com.ai.deepsearch</groupId>
<artifactId>ImageSearchEngine</artifactId>
<version>1.0-SNAPSHOT</version>
<packaging>war</packaging>

<name>ImageSearchEngine Maven Webapp</name>

<properties>
  <project.build.sourceEncoding>UTF-8</project.build.sourceEncoding>
  <maven.compiler.source>1.7</maven.compiler.source>
  <maven.compiler.target>1.7</maven.compiler.target>
  <!-- Change the nd4j.backend property to nd4j-cuda-8.0-platform to
use CUDA GPUs -->
  <nd4j.backend>nd4j-native-platform</nd4j.backend>
  <!--<nd4j.backend>nd4j-cuda-8.0-platform</nd4j.backend>-->
  <java.version>1.8</java.version>
  <nd4j.version>0.9.1</nd4j.version>
  <dl4j.version>0.9.1</dl4j.version>
  <datavec.version>0.9.1</datavec.version>
  <arbiter.version>0.9.1</arbiter.version>
  <logback.version>1.1.7</logback.version>
  <mapdb.version>3.0.6</mapdb.version>
  <tarsosLSH.version>1.0</tarsosLSH.version>
</properties>

<dependencyManagement>
  <dependencies>
    <dependency>
      <groupId>org.nd4j</groupId>
      <artifactId>nd4j-native-platform</artifactId>
```

```xml
      <version>${nd4j.version}</version>
    </dependency>
    <dependency>
      <groupId>org.nd4j</groupId>
      <artifactId>nd4j-cuda-8.0-platform</artifactId>
      <version>${nd4j.version}</version>
    </dependency>
  </dependencies>
</dependencyManagement>

<dependencies>
  <dependency>
    <groupId>junit</groupId>
    <artifactId>junit</artifactId>
    <version>4.11</version>
    <scope>test</scope>
  </dependency>
  <dependency>
    <groupId>javax.servlet</groupId>
    <artifactId>javax.servlet-api</artifactId>
    <version>3.1.0</version>
    <scope>provided</scope>
  </dependency>
  <dependency>
    <groupId>commons-fileupload</groupId>
    <artifactId>commons-fileupload</artifactId>
    <version>1.3.3</version>
  </dependency>
  <dependency>
    <groupId>commons-io</groupId>
    <artifactId>commons-io</artifactId>
    <version>2.2</version>
```

```xml
    </dependency>
    <dependency>
        <groupId>org.nd4j</groupId>
        <artifactId>${nd4j.backend}</artifactId>
    </dependency>
    <dependency>
        <groupId>org.deeplearning4j</groupId>
        <artifactId>deeplearning4j-core</artifactId>
        <version>${dl4j.version}</version>
    </dependency>
    <dependency>
        <groupId>org.deeplearning4j</groupId>
        <artifactId>arbiter-deeplearning4j</artifactId>
        <version>${arbiter.version}</version>
    </dependency>
    <dependency>
        <groupId>ch.qos.logback</groupId>
        <artifactId>logback-classic</artifactId>
        <version>${logback.version}</version>
    </dependency>
    <dependency>
        <groupId>org.mapdb</groupId>
        <artifactId>mapdb</artifactId>
        <version>${mapdb.version}</version>
    </dependency>
    <dependency>
        <groupId>be.tarsos</groupId>
        <artifactId>TarsosLSH</artifactId>
        <version>${tarsosLSH.version}</version>
    </dependency>
</dependencies>
```

```xml
<build>
  <finalName>ImageSearchEngine</finalName>
  <pluginManagement><!-- lock down plugins versions to avoid using
Maven defaults (may be moved to parent pom) -->
    <plugins>
      <plugin>
        <artifactId>maven-clean-plugin</artifactId>
        <version>3.0.0</version>
      </plugin>
      <!-- see http://maven.apache.org/ref/current/maven-core/default-
bindings.html#Plugin_bindings_for_war_packaging -->
      <plugin>
        <artifactId>maven-resources-plugin</artifactId>
        <version>3.0.2</version>
      </plugin>
      <plugin>
        <artifactId>maven-compiler-plugin</artifactId>
        <version>3.7.0</version>
      </plugin>
      <plugin>
        <artifactId>maven-surefire-plugin</artifactId>
        <version>2.20.1</version>
      </plugin>
      <plugin>
        <artifactId>maven-war-plugin</artifactId>
        <version>3.2.0</version>
      </plugin>
      <plugin>
        <artifactId>maven-install-plugin</artifactId>
        <version>2.5.2</version>
      </plugin>
      <plugin>
```

```
        <artifactId>maven-deploy-plugin</artifactId>
        <version>2.8.2</version>
      </plugin>
    </plugins>
  </pluginManagement>
</build>
</project>
```

與前一個子專案的 pom.xml 檔案大致相同。首先是子專案的基礎資訊 groupId、artifactId、version，以及包裝方式 packaging 和專案名稱 name。接下來 properties 部分定義了編譯器的版本、nd4j 函數庫的後端選擇 nd4j.backend，以及所有需要引用依賴的版本。下面的 dependencyManagement 部分，對 nd4j 函數庫需要的兩種後端 nd4j-native-platform 和 nd4j-cuda-8.0- platform 進行了宣告，便於我們在使用時根據需要進行引用。dependencies 部分定義了需要引用的各種依賴：servlet 函數庫、apache 的上傳元件套件 commons-fileupload、nd4j 的後端（我們這裡只使用 CPU 而不使用 GPU，所以引用的是 nd4j-native-platform）、DeepLearning4J 的核心元件 deeplearning4j-core、DeepLearning4J 中用於超參數最佳化的 arbiter-deeplearning4j、DeepLearning4J 中用於記錄檔系統的 logback、嵌入式資料庫 mapdb、用於實現 LSH 演算法的 TarsosLSH。最後的 build 部分中定義了最後產生的專案名稱，以及各種編譯、包裝、安裝、部署、清理、資源等外掛程式。

由於 TarsosLSH 的作者並沒有將其發佈到遠端中心倉庫中，所以我們需要將其安裝到本機倉庫中，以便 Maven 能夠自動引用該依賴。首先，造訪 https://0110.be/releases/TarsosLSH/ TarsosLSH-latest/ 下載 TarsosLSH-latest.jar 套件。如圖 5-11 所示，接下來進入選單，依次選

擇 "Run → Edit Configurations" 設定安裝 TarsosLSH-latest.jar 套件的指令。然後在 Run/Debug Configurations 對話方塊中點擊左上角的綠色加號，在 Add New Configuration 下選擇 "Maven"。而後在對話方塊右側的 Name 中填入 "install_tarsosLSH"，在 Command line 中填入以下指令：install: install-file-Dfile=X:\xxx\TarsosLSH-latest.jar-DgroupId=be.tarsos-DartifactId= TarsosLSH-Dversion=1.0-Dpackaging=jar。其中，Dfile 代表 TarsosLSH-latest.jar 儲存的實際位置，DgroupId、DartifactId、Dversion 分別代表 TarsosLSH-latest.jar 的 groupId、artifactId 和 version，Dpackaging = jar 代表包裝形式為 jar。

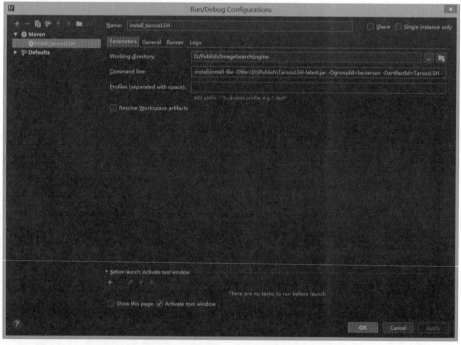

圖 5-11 安裝 TarsosLSH-latest.jar 到本機 Maven 倉庫的 Run 設定

當以上 install_tarsosLSH 執行設定設定好後，便可以透過執行 install_tarsosLSH 指令，將 TarsosLSH-latest.jar 安裝到 Maven 的本機函數庫中。接下來可以看到當 install_tarsosLSH 指令成功執行後，原本紅色的 TarsosLSH 依賴版本部分 "<version>${tarsosLSH.version}</version>" 恢復正常，表示該依賴已經成功引用。

在引用相關依賴項以後，就可以編輯程式實現影像搜尋引擎子專案的各項功能了。由於在該子專案中並不需要使用 jsp，所以 maven-archetype-webapp 原型範本在 src\main\webapp\ 目錄下建立的 index.jsp 檔案可以直接刪除，取而代之的是我們接下來建立的 index.html。該子專案中的 Servlet-api 採用 3.0 以上版本，不需要在 web.xml 中對 Servlet 進行設定，web.xml 檔案只指明了採用 index.html 作為歡迎頁，相當簡潔，程式如下。

▌ 程式 5-5：web.xml

```
<!DOCTYPE web-app PUBLIC
 "-//Sun Microsystems, Inc.//DTD Web Application 2.3//EN"
 "http://java.sun.com/dtd/web-app_2_3.dtd" >

<web-app>
  <display-name>Archetype Created Web Application</display-name>
  <welcome-file-list>
    <welcome-file>index.html</welcome-file>
  </welcome-file-list>
</web-app>
```

𝕏 程式 5-6：index.html

```html
<html>
<head>
    <meta http-equiv="Content-Type" content="text/html; charset=utf-8" />
    <title>影像搜索</title>
    <meta name="renderer" content="webkit">
    <meta http-equiv="X-UA-Compatible" content="IE=Edge,chrome=1">
    <link rel="stylesheet" type="text/css" href="./css/style.css" />
    <script src="./js/jquery-1.9.1.min.js" type="text/javascript">
</script>
</head>

<body>
<div class="logo_box"><img src="./img/logo.png"></div>
<div class="search_box">
    <form id="file_upload" enctype="multipart/form-data" action="search"
method="post">
        <input type="text" class="search_input" maxlength="100">
        <div class="search_btn">搜索</div>
        <div class="img_camera"></div>
        <img class="img_preview" src="" style="display: none;"/>
        <div class="img_camera_mouseenter" style="display: none;">選擇圖
片後點擊搜索</div>
        <i class="img_camera_mouseenter_arrow" style="display: none;">
</i>
        <input type="file" name="file_choice" id="file">
    </form>
</div>
<script type="text/javascript" src="./js/upload.js" crossorigin=""></script>
</script>
</body>
</html>
```

▓ 程式 5-7：upload.js

```javascript
$(".img_camera").on({
    click: function () {
        $(".img_camera_mouseenter,.img_camera_mouseenter_arrow").hide();
        $("#file").click();
    },
    mouseenter: function () {
        timer = setTimeout(function () {
            $(".img_camera_mouseenter,.img_camera_mouseenter_arrow").
slideDown(200);
        }, 200);
    },
    mouseleave: function () {
        clearTimeout(timer);
        $(".img_camera_mouseenter,.img_camera_mouseenter_arrow").
slideUp(200);
    }
});


$("#file").on("change", function (event) {
    var file = event.target.files || e.dataTransfer.files;
    if (file) {
        var fileReader = new FileReader();
        fileReader.onload = function () {
            $(".img_preview").attr("src", this.result);
        }
        fileReader.readAsDataURL(file[0]);
        $(".img_preview").show();

    }
});
```

```
$(".search_btn").on({
    click: function () {
        $("#file_upload").submit();
    }
});
```

在程式 5-6 和程式 5-7 中，index.html 透過 form 表單向名為 search 的相對 URL 傳送查詢圖片。upload.js 則呼叫 JQuery 函數庫完成了「搜索」按鈕點擊觸發 form 表單的傳送事件，並實現了一些簡單的互動效果。

除此以外，我們還需要建立 WebEngineInit、SearchImageServlet、SearchSimilarImgs、Utils 這 4 個類別，分別實現 VGG16 模型預先載入、搜索請求處理與回饋、線性和局部敏感雜湊方法尋找相似影像、工具類別等功能。下面來對它們逐一說明：

⧗ **程式 5-8：WebEngineInit.java**

```java
package com.ai.deepsearch;

import org.deeplearning4j.nn.graph.ComputationGraph;
import org.deeplearning4j.util.ModelSerializer;

import javax.servlet.ServletContextEvent;
import javax.servlet.ServletContextListener;
import javax.servlet.annotation.WebListener;
import java.io.File;
import java.io.IOException;

/**
 *  引擎初始化
 */
```

```java
@WebListener
public class WebEngineInit implements ServletContextListener {
    public static ComputationGraph vgg16Model;

    public void contextInitialized(ServletContextEvent event) {
        System.out.println("Web initialized");
        try {
            String modelFilePath=WebEngineInit.class.getClassLoader().
getResource ("vgg16.zip").getPath().substring(1);
            File vgg16ModelFile=new File(modelFilePath);
            vgg16Model= ModelSerializer.restoreComputationGraph
(vgg16ModelFile);
        } catch (IOException e) {
            e.printStackTrace();
        }
    }

    public void contextDestroyed(ServletContextEvent event) {
        System.out.println("Web destroyed");
        vgg16Model=null;
    }
}
```

在程式 5-8 中，WebEngineInit 類別實現了 ServletContextListener 介面，使其能夠監聽 ServletContext 物件的生命週期。而 Servlet 的容器啟動或終止應用時會觸發 ServletContextEvent 事件，這樣我們就可以在 Tomcat 啟動時做一些耗時的初始化工作，在 Tomcat 終止應用時做一些資料的清理工作。當 Tomcat 啟動時，我們在 contextInitialized 方法中，載入了在 ImageNet 影像函數庫上預訓練的 VGGNet16 模型檔案 vgg16.zip。

⏳ 程式 5-9：SearchImageServlet.java

```java
package com.ai.deepsearch;

import org.apache.commons.fileupload.FileItem;
import org.apache.commons.fileupload.FileUploadException;
import org.apache.commons.fileupload.disk.DiskFileItemFactory;
import org.apache.commons.fileupload.servlet.ServletFileUpload;
import org.datavec.image.loader.NativeImageLoader;
import org.nd4j.linalg.api.ndarray.INDArray;
import org.nd4j.linalg.dataset.api.preprocessor.DataNormalization;
import org.nd4j.linalg.dataset.api.preprocessor.VGG16ImagePreProcessor;
import javax.servlet.ServletException;
import javax.servlet.annotation.WebServlet;
import javax.servlet.http.HttpServlet;
import javax.servlet.http.HttpServletRequest;
import javax.servlet.http.HttpServletResponse;
import java.io.File;
import java.io.IOException;
import java.io.PrintWriter;
import java.util.List;
import java.util.Map;
import java.util.Set;

/**
 *   搜索請求 Servlet
 */
@WebServlet(value = "/search")
public class SearchImageServlet extends HttpServlet {

    @Override
    public void doPost(HttpServletRequest request,
HttpServletResponse response)
```

```java
throws ServletException, IOException {
        System.out.println("SearchImageServlet servlet handling post");
        try {
            DiskFileItemFactory factory = new DiskFileItemFactory();
            File f=new File("E:\\storetest");
            factory.setRepository(f);
            ServletFileUpload fileUpload = new
ServletFileUpload(factory);
            String uploadDir = request.getSession().getServletContext().
getRealPath  ("/upload_imgs");
            List<FileItem> fileItems = fileUpload.parseRequest(request);
            System.out.println("file items size:"+fileItems.size());
            for (FileItem item : fileItems) {
                if(!item.isFormField()) {
                    String fileName=item.getName();
                    if(fileName.lastIndexOf("\\")>=0) {
                        fileName=fileName.substring(fileName.
lastIndexOf("\\"));
                    } else {
                        fileName=fileName.substring(fileName.
lastIndexOf("\\")+1);
                    }
                    File uploadFile=new File(uploadDir+"/"+fileName);
                    if(!uploadFile.exists()) {
                        uploadFile.getParentFile().mkdirs();
                    }
                    uploadFile.createNewFile();
                    item.write(uploadFile);
                    item.delete();

                    NativeImageLoader loader = new
NativeImageLoader(224,224,3);
```

```
                    INDArray imageArray=loader.asMatrix(uploadFile);
                    DataNormalization scaler = new
VGG16ImagePreProcessor();
                    scaler.transform(imageArray);

                    Map<String,INDArray> map=WebEngineInit.vgg16Model.
feedForward
(imageArray,false);
                    INDArray feature=map.get("fc2");

                    String imagesDb = request.getSession().
getServletContext().getRealPath("/WEB-INF/images.db");

                    Set<String> result=SearchSimilarImgs.search(true,
imagesDb, fileName, feature);

                    response.setContentType("text/html;charset=utf-8");
                    PrintWriter writer=response.getWriter();
                    writer.println("<html>");
                    writer.println("<head>");
                    writer.println("<title> 查詢結果 </title>");
                    writer.println("<link rel=\"stylesheet\"
type=\"text/css\" href=\"./css/style.css\" />");
                    writer.println("</head>");
                    writer.println("<body>");
                    writer.println("<div class=\"search\">");
                    writer.println("<div class=\"back_btn\"><a href =
\"http://localhost:8080/imgsearch/\"> 回首頁 </a></div>");
                    writer.println("<div class=\"title_search\"><span>
查詢影像 </span> </div>");
                    writer.println("</div>");
                    writer.println("<div>");
```

```
                    writer.println("<div class=\"search_img\"><img src=
\"./upload_imgs/"+fileName+"\"></div>");
                    writer.println("</div>");
                    writer.println("<div class=\"line\"></div>");
                    writer.println("<div class=\"simi\">");
                    writer.println("<div class=\"title_simi\"><span>
相似影像</span> </div>");
                    writer.println("</div>");
                    writer.println("<div id=\"result\">");
                    for(String r:result) {
                        writer.println("<div class=\"simi_img\"><img
src=\"./image/"+r+"\"></div>");
                        System.out.println(r);
                    }
                    writer.println("</div>");
                    writer.println("</body>");
                    writer.println("</html>");
                    writer.flush();
                    writer.close();
                }
            }
        } catch (FileUploadException e) {
            e.printStackTrace();
        } catch (Exception e) { // item.write(uploadImages)
            e.printStackTrace();
        }
    }
}
```

在 程 式 5-9 中，SearchImageServlet 類 別 是 一 個 將 前 端 頁 面 和 後
端 邏 輯 緊 密 連 接 起 來 的 Servlet。 它 透 過 使 用 apache 的 commons-
fileupload 元 件 接 收 和 處 理 index.html 中 form 表 單 傳 送 到 search 處

的影像。然後 SearchImageServlet 將該影像進行一些前置處理（包含 NativeImageLoader 和 VGG16ImagePreProcessor），並分析該影像的 VGG16 模型的 FC2 層特徵。接下來將該特徵送入 SearchSimilarImgs 類別搜索相似影像，並將傳回結構以 html 的形式回饋給使用者。

⚡ 程式 5-10：SearchSimilarImgs.java

```java
package com.ai.deepsearch;

import be.tarsos.lsh.LSH;
import be.tarsos.lsh.Vector;
import be.tarsos.lsh.families.CosineHashFamily;
import be.tarsos.lsh.families.DistanceMeasure;
import be.tarsos.lsh.families.EuclideanDistance;
import be.tarsos.lsh.families.HashFamily;
import org.mapdb.DB;
import org.mapdb.DBMaker;
import org.mapdb.Serializer;
import org.nd4j.linalg.api.ndarray.INDArray;

import java.util.ArrayList;
import java.util.LinkedHashSet;
import java.util.List;
import java.util.Set;
import java.util.concurrent.ConcurrentMap;

/**
 * 尋找相似影像
 */
public class SearchSimilarImgs {

    public static List<Vector> getVectorListFromDB(String dbPath) {
        DB db = DBMaker.fileDB(dbPath).make();
```

```
        ConcurrentMap<String, double[]> map = db.hashMap("feat_map",
Serializer.STRING, Serializer.DOUBLE_ARRAY).open();
        //int size=map.size();
        List<Vector> vecs = new ArrayList<Vector>();
        for (String key : map.keySet()) {
            //int dimension=map.get(key).length;
            double[] val = map.get(key);
            // norm2
            val = Utils.normalizeL2(val);
            Vector vec = new Vector(key, val);
            vecs.add(vec);
        }
        db.close();
        return vecs;
    }


    public static Set<String> search(boolean linear, String dbPath,
String imgName, INDArray fc2Feat) {
        Set<String> similarImgsName = new LinkedHashSet<String>();
        List<Vector> dataset = getVectorListFromDB(dbPath);
        int dimension = dataset.get(0).getDimensions();
        HashFamily cosHashFamily = new CosineHashFamily(dimension);

        double[] queryFeat = Utils.INDArray2DoubleArray(fc2Feat);
        // norm2
        queryFeat = Utils.normalizeL2(queryFeat);
        Vector queryVec = new Vector(imgName, queryFeat);
        int numOfNeighbours = 5;
        DistanceMeasure disMeasure = new EuclideanDistance();

        if (linear) {
            List<Vector> neighbours = LSH.linearSearch(dataset,
```

```
queryVec, numOfNeighbours, disMeasure);
            for (Vector neighbour : neighbours) {
                similarImgsName.add(neighbour.getKey());
            }
        } else {
            LSH lsh = new LSH(dataset, cosHashFamily);
            int numOfHashes = 3;
            int numOfHashTables = 2;
            lsh.buildIndex(numOfHashes, numOfHashTables);

            List<Vector> neighbours = lsh.query(queryVec,
numOfNeighbours);
            for (Vector neighbour : neighbours) {
                similarImgsName.add(neighbour.getKey());
            }
        }

        return similarImgsName;
    }
}
```

在程式 5-10 中，SearchSimilarImgs 類別包含 getVectorListFromDB 和
search 兩個函數。getVectorListFromDB 將特徵索引函數庫 images.db 中
儲存的全部（影像名、特徵碼）鍵值對取出。而 images.db 正是由我們
前面建立的特徵索引函數庫產生工具 GenerateImgsFeatDBTool.jar 所產
生的。如圖 5-12 所示，在命令列狀態下輸入：

```
java -jar GenerateImgsFeatDBTool.jar
    -m 專案路徑 \src\main\resources\vgg16.zip
    -d 專案路徑 \src\main\webapp\WEB-INF\images.db
    -i 專案路徑 \src\main\webapp\image
```

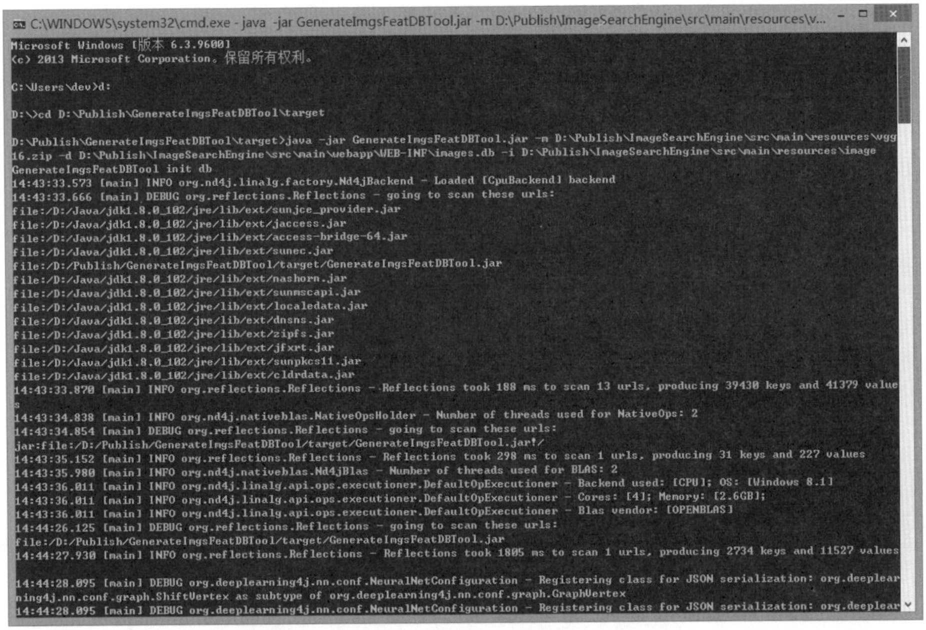

圖 5-12　產生特徵索引函數庫 images.db

將在目錄「專案路徑 \src\main\webapp\WEB-INF\」下產生特徵索引函
數庫 images.db。search 函數依據使用者是否採用線性方法，來選擇將
查詢影像的特徵與特徵索引函數庫一一比較，還是使用局部敏感雜湊
（LSH）做候選特徵集內的局部比較。

⊠ 程式 5-11：Utils.java

```java
package com.ai.deepsearch;


import org.nd4j.linalg.api.ndarray.INDArray;


import java.util.Arrays;


/**
 *  工具類別
```

```
    */
public class Utils {
    public static double[] INDArray2DoubleArray(INDArray indArr) {
        String indArrStr = indArr.toString().replace("[", "").
replace("]", "");
        String[] strArr = indArrStr.split(",");
        int len = strArr.length;
        double[] doubleArr = new double[len];
        for (int i = 0; i < len; i++) {
            doubleArr[i] = Double.parseDouble(strArr[i]);
        }
        return doubleArr;
    }

    public static double[] normalizeL2(double[] vec) {
        double norm2 = 0;
        for (int i = 0; i < vec.length; i++) {
            norm2 += vec[i] * vec[i];
        }
        norm2 = (double) Math.sqrt(norm2);
        if (norm2 == 0) {
            Arrays.fill(vec, 1);
        } else {
            for (int i = 0; i < vec.length; i++) {
                vec[i] = vec[i] / norm2;
            }
        }
        return vec;
    }
}
```

在程式 5-11 中，Utils 類別是一個工具類別，它提供 INDArray2DoubleArray
和 normalizeL2 兩個函數，前者能夠將 INDArray 類型轉為 Double[] 類
型，後者對向量進行歸一化。

至此，影像搜尋引擎的子專案已全部實現。下面我們將要把它部署到
tomcat 上測試，看一下效果如何。由於 IntelliJ IDEA 社區版並沒有提供
設定 tomcat 的功能，這裡需要使用 Maven 的 tomcat 外掛程式來進行測
試。點擊選單，依次選擇 "Run → Edit Configurations"，在出現的 Run/
Debug Configurations 對話方塊中進行執行設定。點擊左上角的綠色加
號，在 Add New Configuration 下選擇 "Maven"，建立 Name 為 "tomcat"
的指令。如圖 5-13 所示，在 Command line 中填入指令 "tomcat7:run"。
此外，還需要編輯 pom.xml 下載設定對應的 tomcat 外掛程式，程式如
下。

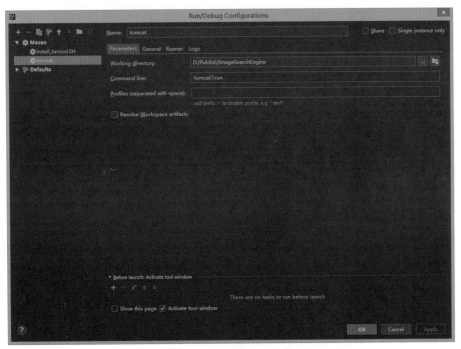

圖 5-13 tomcat 設定

⏳ **程式 5-12：pom.xml 中的 tomcat 外掛程式部分**

```
<plugin>
  <groupId>org.apache.tomcat.maven</groupId>
  <artifactId>tomcat7-maven-plugin</artifactId>
  <version>2.1</version>
  <configuration>
    <port>8080</port>
    <path>/imgsearch</path>
    <uriEncoding>UTF-8</uriEncoding>
    <server>tomcat7</server>
  </configuration>
</plugin>
```

在一切準備妥當之後，點擊選單，依次選擇 "Run → Run tomcat" 便可以將該子專案部署到 tomcat 上。在 tomcat7-maven-plugin 外掛程式下載各種原始程式碼和資源，編譯複製完成後，開啟瀏覽器，在位址框中輸入 "http://localhost:8080/imgsearch/" 便可以進入 index.html 首頁，如圖 5-14 所示。

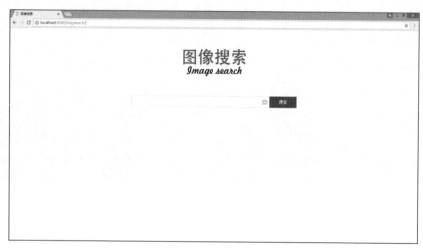

圖 5-14 影像搜尋引擎首頁

上傳查詢圖片後，搜索框中會顯示它的縮圖。點擊「搜索」按鈕，等待片刻，便會看到系統傳回的結果，如圖 5-15 所示。

圖 5-15　影像搜尋引擎傳回相似影像

5.3 最佳化策略

看到圖 5-15 中影像搜尋引擎傳回的相似影像，讀者可能會有這樣的疑問：系統傳回的相似影像和查詢影像不是很像啊？透過仔細觀察，我們會發現這些衣服的花紋是近似的。之所以款式有差別，是因為我們使用 GenerateImgsFeatDBTool 工具產生的特徵索引函數庫，只是採用 400 餘幅影像的資料集產生的（如圖 5-16 所示），並且資料集中沒有這種款式的長裙。

圖 5-16　用於產生特徵索引函數庫的圖像資料集

那麼我們怎樣才能最佳化該引擎，進一步解決這個問題呢？首先根據該引擎的工作需求，儘量擴充用於建置特徵索引函數庫的資料集的規模和覆蓋面，使其有一定的廣度和深度，這樣才能使該引擎傳回的結果更加符合相似性的要求。其次，該引擎影像特徵分析功能以在 ImageNet 資料集上預訓練為基礎的 VGG16 模型實現，和目標資料有一定差異。為此，我們可以搜集一定量的垂直領域資料集來微調模型，使分析的特徵更加符合該領域的特點，進一步傳回更高品質的結果。

5.4 本章小結

本章帶領讀者使用前面各個章節說明的內容,從零開始建置一個線上影像搜尋引擎。無論是專案架構設計的說明、開發技術路線的選擇、開發環境的設定和使用,還是實際的程式實現,本章都給予詳細的說明。透過對本章的學習,讀者已能夠透徹地了解影像檢索的理論,並具有獨立實現一個 Web 影像搜尋引擎的實際能力。最後作者指出該影像搜尋引擎進一步改進和最佳化的策略和方向,為讀者提供了結合本身需求進一步改進該專案的空間。

Note

Note

Note

Note